The Economics of Biodiversity and Ecosystem Services

Ecosystems and biodiversity have been degraded over decades due to human activities. One of the critical causes is market failure: the current market only accounts tangible resources and neglects intangible functions, such as climate control and natural hazard mitigation. Under such circumstances in capitalism, land conversion and resource exploitation, which generate financial income, are highly prioritized over conservation, which is not necessarily beneficial in monetary terms.

To halt ecosystem degradation, thus, the values of ecosystem services need to be visualized and economic instruments for ecosystem conservation should be further developed. This book focuses on these two aspects and performs several studies, including valuation of ecosystem services, productivity analysis, institutional design of payment for ecosystem services (PES), impact assessment of reduction emission from deforestation and forest degradation (REDD), and economic experiment of mitigation banking scheme. From these analyses, economic values of ecosystem services are demonstrated from both the supply and demand sides, and the directions for improving economic instruments are indicated both directly and indirectly.

As many of these analyses are usually conducted in North America and Europe, this book is unique in geographical focus, namely, Japan, Asia, and the globe. Also, a wide variety of ecosystems are targeted for studies, such as agricultural lands, forests, wetlands, and marine environments. Hence, this will be an informative introduction for those who desire to study the economics of biodiversity and ecosystem services in these regions and of these ecological zones.

Shunsuke Managi is the Associate Professor of Resource and Environmental Economics at the Tohoku University, Japan, while also holding a position as an Adjunct Associate Professor at the University of Tokyo. He is an editor of *Environmental Economic and Policy Studies*, a lead author for the Intergovernmental Panel on Climate Change, and is the author of *Technology, Natural Resources and Economic Growth: Improving the Environment for a Greener Future*, published by Edward Elgar Publishing Ltd. in 2011. He is the author of 8 books and 100 academic journal papers.

Routledge Explorations in Environmental Economics
Edited by Nick Hanley
University of Stirling, UK

The Economics of Biodiversity and Ecosystem Services

Edited by Shunsuke Managi

Routledge
Taylor & Francis Group

LONDON AND NEW YORK

First published 2013
by Routledge
2 Park Square, Milton Park, Abingdon, Oxon, OX14 4RN

Simultaneously published in the USA and Canada
by Routledge
711 Third Avenue, New York, NY 10017

Routledge is an imprint of the Taylor & Francis Group, an informa business

British Library Cataloguing in Publication Data
A catalogue record for this book is available from the British Library

Library of Congress Cataloging in Publication Data
The economics of biodiversity and ecosystem services/edited by
Shunsuke Managi.
    p. cm. – (Routledge explorations in environmental economics; 38)
    Includes bibliographical references and index.
    1. Ecosystem services.
    2. Biodiversity–Economic aspects.
    3. Ecosystem management.
    I. Managi, Shunsuke.
QH541.15.E267E27 2012
333.95'16–dc23   2012008981

ISBN: 978-0-415-62563-0 (hbk)
ISBN: 978-0-203-09787-8 (ebk)

Typeset in Times New Roman
by Sunrise Setting Ltd, Torquay, UK

Printed and bound in the United States of America by Publishers Graphics,
LLC on sustainably sourced paper.

Contents

Figures

Tables

Preface

Why do we need to conserve biodiversity? How can we effectively conserve so as not to lose it? Humans benefit from ecosystem services, both directly and indirectly. Likewise, economic activities are based on these services. However, ecosystems cannot provide services as they used to do when they have been changed due to the great impacts of human and economic activities. The direct drivers of change include habitat change, climate change, invasive alien species, over-exploitation, and pollution, backed by indirect factors such as population change, change in economic activities, socio-political factors, cultural factors, and technological change. In brief, ecosystem degradation is considered to be an economic problem.

Ecologically, there are no ideal states for nature. Conservation of the ecological system – the components of biological nature – stands for biodiversity conservation. There are two main types of value in biodiversity: one is that derived from the protection of usable resources, while the other is the hope of its preservation now, because current human knowledge cannot judge it as valueless. Value itself is determined by a social consensus based on human value judgments. As both the factors influencing biodiversity and the value judgments thereof are relevant to economics, analysis and measures based on economics are required.

The Convention of Biological Diversity (CBD) with the aims of the conservation of biological diversity, the sustainable use of its components, as well as the fair and equitable sharing of the benefits arising out of the utilization of genetic resources was adopted in 1992. It came into force at the end of 1993. At the Sixth Conference of the Parties (COP) in 2002, the Strategic Plan for the CBD was adopted and the 2010 Biodiversity Target – to achieve a significant reduction of the current rate of biodiversity loss by 2010 – was framed. The COP10 was held at Nagoya, Aichi Prefecture, in October 2010. The new strategic plan, the Aichi Target, to "take effective and urgent action to halt the loss of biodiversity in order to ensure that by 2020 ecosystems are resilient and continue to provide essential services" was adopted as the post-2010 Biodiversity Target. Simultaneously, the Nagoya Protocol relevant to access and benefit-sharing (ABS) of genetic resources was also adopted as the subsequent protocol to the Cartagena Protocol (adopted in 2000; it came into force in 2003).

In response to these international actions, the National Strategy for the Conservation and Sustainable Use of Biological Diversity based on the CBD was formulated in 2005 in Japan. Likewise, the Basic Act on Biodiversity came into force in 2008. On the basis of this legal background the Japanese government plans to implement various measures for biodiversity conservation.

Considering such importance, the focus of this book on studies on the evaluation and institutional analysis of ecosystem management is mainly from an economic perspective. Our fundamental objective is to contribute to the sustainable use of ecosystem services at the national and global levels. Traditionally, domestic biodiversity has been evaluated and analyzed mainly from ecological viewpoints; political and economic approaches have not been emphasized as much. We expected that we could evaluate various institutions from economic perspectives by revealing the economic effects of management systems, and thereby contribute to framing future resource-management systems.

The whole structure of institutions is accurately comprehended in this book through the quantitative analysis of ecosystem services at the macro level. Furthermore, economic analysis is conducted so as to enhance the accuracy of basic analysis, and the policy suggestions made are directed towards local entities.

This book compiles results from the study conducted under the commissioned project of the Ministry of Environment Japan: "Policy studies on environmental economics contributing to the world: Study on the policy options aiming for sustainable use of ecosystem services through internalization of economic benefits," as well as the Grant-in-aid for Scientific Research from the Ministry of Education, Culture, Sports, Science, and Technology. We would like to express our sincere gratitude to those who kindly provided these research opportunities. We also would like to thank those who attended our conference presentation – as well as our regular research group – for their discussion. Finally, we appreciate the kindness of both domestic and overseas policy makers as well as stakeholders who kindly enabled us to conduct our field research.

Shunsuke Managi
1 June 2012

Acknowledgments

Most of the analyses in this book were financially supported by the commissioned project of Ministry of Environment Japan, "Policy studies on environmental economics contributing to the world: study on the policy options aiming for sustainable use of ecosystem services through internalization of economic benefits." We would like to express our sincere appreciation.

Part I
Biodiversity and ecosystem services

1 Sustainable use of ecosystem services

Kei Kabaya and Shunsuke Managi

Ecosystem services

Ecosystem and biodiversity

Since the conclusion of the Convention on Biological Diversity (CBD) at the 1992 United Nations Conference on Environment and Development (UNCED) (the so-called Earth Summit) in Rio de Janeiro, Brazil, biodiversity has been attracting global attention to the point where its loss has become one of the most important environmental issues in the world. Climate change has followed a similar pathway (see Figure 1.1) but these two issues have different characteristics.

Climate change is an internationally common issue triggered by a single cause, i.e. greenhouse gases (GHG), and this can be dealt with by setting an emissions-reduction target in each nation. On the other hand, loss of biodiversity is a multi-layered issue caused by various factors at local, national, regional, and global levels. The countermeasures vary widely from gene preservation to species conservation and protected-area designation. The difference between the two issues could be simply illustrated with the example of plantations: tree species which grow faster are preferable from the perspective of combating climate change, while tree species should be limited to native ones from the viewpoint of biodiversity conservation. In short, biodiversity, due to its complexity, is more difficult to understand compared to climate change.

Accurate recognition of its definition will help to deepen the understanding of complicated biodiversity. It is defined in the CBD Article 2 as "the variability among living organisms from all sources including, inter alia, terrestrial, marine and other aquatic ecosystems and the ecological complexes of which they are part; this includes diversity within species, between species and of ecosystems" (UN 1992). In essence, biodiversity means the diversity of genes, species, and ecosystems; more precisely, biodiversity is the expression of the diverse "state" of those. On the other hand, an ecosystem, which tends to be recognized as similar to biodiversity, is defined as "a dynamic complex of plant, animal, and microorganism communities and the nonliving environment interacting as a functional unit" (MA 2005a: 8). In short, an ecosystem is the "function," and diversity of ecosystems is a part of biodiversity.

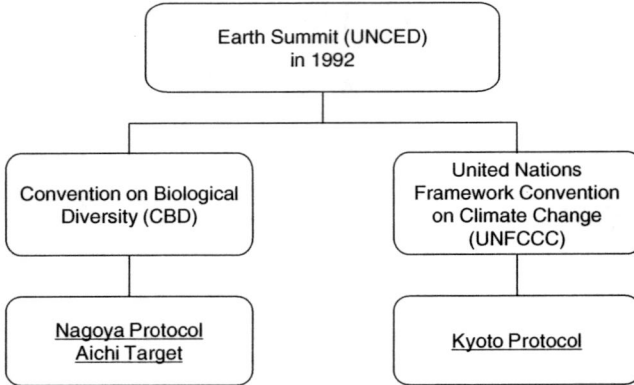

Figure 1.1 International frameworks for biodiversity and climate change.

Biodiversity tends to provide people with the image of species diversity, because the concept of species is relatively clearly defined and is familiar to us. Although the difference between an African elephant (*Loxodonta Africana*) and an Asian elephant (*Elephas maximus*) is difficult to observe, for instance, an elephant can be easily distinguished from a rhinoceros. Likewise, the difference between an oak and a pine can be easily recognized, while it takes more technical expertise to tell a common oak (*Quercus robur*) from a Japanese oak (*Quercus crispula*). As for the relationship between species and ecosystems, Tilman *et al.* (2005) show that more diverse species generate higher productivity and stability of ecosystems, and Elmqvist *et al.* (2003) demonstrate that species diversity contributes to the enhancement of resilience in the face of disturbance. These relationships are not necessarily linear; however, upward rigidity is pointed out through the rivet hypothesis (Ehrlich and Ehrlich 1981). This hypothesis likens an ecosystem to an airplane, which can continue its flight even if some rivets drop off, but cannot do so when a certain number of rivets are lost. Similarly, an ecosystem can maintain its function when some species die off, but its function will be disrupted when a certain number of species become extinct. This may imply that not all the species need to be conserved to maintain the functions of ecosystems, however, it is still not clear which species have a minimum effect on the functions and, thus, can be erased. Taking into consideration that a substantial number of species have not yet been recorded and that the relationships among species are too complicated to comprehend all at once with current human knowledge, it is obvious that we should not easily accept the extinction of even one species.

The component of biodiversity which we should prioritize is a controversial issue. Genetic resources are important sources for development of new medicines, but it will be impossible to preserve all of the genes, which may differ from individual to individual. As such, conservation of tens of millions to a hundred million species is very tough work under the current budgetary conditions in the world. It will be possible to focus on which specific keystone species or umbrella species to conserve, but habitats for those species also need to be conserved in that case. Judging from these problematic conditions and taking into consideration that species and genes cannot be fully protected without conserving ecosystems, priority should be assigned to ecosystems even if they may not be suitable policy targets due to their complexity and difficulties in clear demarcation. As the CBD recognizes the importance of an ecosystem approach, the conservation policies focusing on ecosystems will be the keys for efficient and effective implementation.

Ecosystem services and human well-being

Because ecosystems have complex relationships interactively as well as endemic uniqueness individually, clear and detailed classification is quite a challenging matter. Nonetheless, some attempts to sort ecosystems into several categories have been made, since classification of ecosystems has certain significance for policy-making and conservation activities. The classification by the Millennium Ecosystem Assessment (MA) is introduced as set out below.

First, the components of the Earth are separated into ten systems from their characteristics, and then detailed ecosystems are classified into those systems (see Table 1.1). The areas defined as "permanent water bodies inland from the coastal zone and areas whose properties and use are dominated by the permanent, seasonal, or intermittent occurrence of flooded conditions" are recognized

Table 1.1 Classification of ecosystems

Number	System	Ecosystems included (*examples*)
1	Marine	Continental shelf, slopes, seamounts, and deep sea
2	Coastal	Estuaries, marshes, salt ponds, lagoons, mangroves, deltas, beaches, dunes, coral reefs, seagrass beds, and kelp forests
3	Inland water	Rivers, lakes, floodplains, reservoirs, wetlands, swamps, and inland saline systems
4	Forest	Tropical rainforests, temperate broadleaf forests, mixed forests, and boreal needleleaf forests
5	Dryland	Cultivated lands, scrubland, grassland, semi-desert, and desert
6	Island	Continental islands, volcanic islands, carbonate islands, and archipelagos
7	Mountain	Montane belt, alpine belt, and nival belt
8	Polar	Ice caps, permafrost, tundra, polar deserts, and polar coastal areas
9	Cultivated	Orchards, agro-forestry, and integrated agriculture-aquaculture systems
10	Urban	Residential, commercial, and industrial areas

Source: MA (2005a, b), revised by the authors.

as the inland water system, and ecosystems such as "rivers, lakes, floodplains, reservoirs, wetlands, and inland saline systems" are classified as part of the inland water system (MA 2005a: 30). This categorical classification is considered useful from the perspective of its consistency with the CBD framework and ministerial jurisdictional boundaries regarding primary industries. Note that some terrestrial surfaces and water bodies are classified into two categories or more, e.g. lands for agro-forestry belong to the forest system as well as to the cultivated system.

Ecosystems play substantial roles in maintaining the environment as their original function, such as nutrient cycling, water purification, and air-quality regulation, and in producing natural resources. These are extremely important life-supporting bases for humanity; if these are likened to economic activities, the former can be seen as supply of services and the latter as supply of goods. MA (2005a) calls these services and goods provided by ecosystems "ecosystem services" collectively, and recognizes 31 services within four categories, namely, provisioning, regulating, cultural, and supporting services (see Table 1.2). Ecosystem services vary widely from food provision to climate regulation, recreation and soil formation, and their impacts range broadly from local (e.g. flood regulation) to global (e.g. air quality regulation).

According to MA (2005a), these ecosystem services have deep connections with five elements for human well-being, namely, basic material for good life, health, good social relations, security, and freedom of choice and action. Food, clothing, and shelter represent basic materials for a good life, which greatly depend on the provisioning services of ecosystems. Similarly, nutrients, disease regulation, and mental healthcare are affected by provisioning, regulating, and cultural services. While the share of cultural services based on specific ecosystems may be a basis for good social relations, degradation of provisioning services may cause social conflicts. Security for life and properties is affected by regulating services such as natural-hazard regulation, and freedom of choice and action can be influenced by all of the ecosystem services.

All people are beneficiaries of ecosystem services, but dependency differs in individual situations and in the nature of the services. People in developed countries depend less on wild animals for food sources thanks to highly controlled food production including aquaculture, and they do not need to collect wood as an energy fuel source owing to a sufficient energy supply from fossil fuels. In terms of health and security in developed countries, easy access to medicines and improvement in protective facilities in the case of natural disasters make people less dependent on ecosystem services. On the other hand, poor people in the least developed countries are still highly dependent on wild foods and medicinal plants and are vulnerable to natural disasters due to their living conditions. In short, direct dependency on ecosystem services is greater for the poor than for the rich. Nevertheless, people in developed countries must not neglect the contribution from ecosystem services, considering that foods as natural resources, genetic resources for medicinal development, carbon sequestration and water purification by forests, and ecotourism and scenic beauty are still necessary for their healthy lives.

Table 1.2 Classification of ecosystem services and their current trends

Number	Ecosystem services	Trends*
Provisioning services		
1	Food	
	Crops	▲
	Livestock	▲
	Capture fisheries	▼
	Aquaculture	▲
	Wild plant and animal products	▼
2	Fiber	
	Timber	+/−
	Cotton, hemp, and silk	+/−
3	Fuel	▼
4	Genetic resources	▼
5	Biochemicals, natural medicines, and pharmaceuticals	▼
6	Ornamental resources	N/A
7	Fresh water	▼
Regulating services		
8	Air-quality regulation	▼
9	Climate regulation	
	Global	▲
	Regional and local	▼
10	Water regulation	+/−
11	Erosion regulation	▼
12	Water purification and waste treatment	▼
13	Disease regulation	+/−
14	Pest regulation	▼
15	Pollination	▼
16	Natural-hazard regulation	▼
Cultural services		
17	Cultural diversity	N/A
18	Spiritual and religious values	▼
19	Knowledge systems	N/A
20	Educational values	N/A
21	Inspiration	N/A
22	Aesthetic values	▼
23	Social relations	N/A
24	Sense of place	N/A
25	Cultural-heritage values	N/A
26	Recreation and ecotourism	+/−
Supporting services		
27	Soil formation	†
28	Photosynthesis	†
29	Primary production	†
30	Nutrient cycling	†
31	Water cycling	†

Source: MA (2005a) slightly revised by the authors.
*Legend.
▲: Increasing or enhanced.
▼: Decreasing or degraded.
+/−: Mixed.
NA: Not assessed within the MA.
†: Not applied for supporting services since these services are not directly used by people.

While some ecosystem services can be substituted for by some artificial assets, others cannot. In addition, some of them will incur considerable costs to carry out such substitutions. For instance, livestock can be substituted for wild meats, but it is difficult to substitute livestock for wildlife in ecotourism. Dikes and dams can be substituted for the flood-regulation function by forests and wetlands, although construction of them may place a large burden on local and national accounts. To sum up, degradation of ecosystem services has the possibility to cause irreversible or tremendous damage to society and the economy. Therefore, it would be worth considering giving priority to conservation targets from the perspectives of substitutability and the costs thereof.

Unsustainable use of ecosystem services

Degradation of ecosystems

Ecosystems have been degrading at the most rapid rate ever in the history in the world (MA 2005a). Cumulative deforestation and reclamation mean that cultivated land currently dominates one-quarter of the terrestrial surface, and coastal development in the last few decades has caused the destruction of coral reefs and mangroves by 20 percent and 35 percent, respectively. From three to six times more water resources than of natural river flows are stored in dam sites globally, and species become extinct at a rate of 1,000 times greater than the historical rate. The following examples further clearly describe the severity of ecosystem degradation.

- Tropical rainforests in the island of Borneo, Indonesia, which used to cover almost the whole island, have degraded due to rapid deforestation and development since 1950s. As a result, the total forest areas have almost been halved from their original state (see Figure 1.2) (UNEP/GRID-Arendal 2007).
- The Aral Sea located in Central Asia has lost half of the area of its surface water and two-thirds of its water quantity due to channel modification of the inflowing two rivers for the purpose of upgrading irrigation (see Figure 1.3) (UNEP/GRID-Arendal 2009). It is estimated that this lake will disappear completely by 2020 if the current rate of degradation continues (Pidwirny 1999).
- Despite the fact that Atlantic cod (*Gadus morhua*) fishing off the east coast of Newfoundland has drastically increased its landings since the late 1950s, sufficient conservation activities for fishery stock were not implemented, resulting in the collapse of the fish stock in 1992 and the declaration of an indefinite moratorium on Atlantic cod fishing in 2003 (MA 2005a).

The following five points can be considered as direct drivers for ecosystem degradation, namely, habitat change, climate change, invasive alien species, over-exploitation, and pollution (MA 2005a). Deforestation and channel modification, as above, are examples of habitat changes as well as land conversion, reclamation,

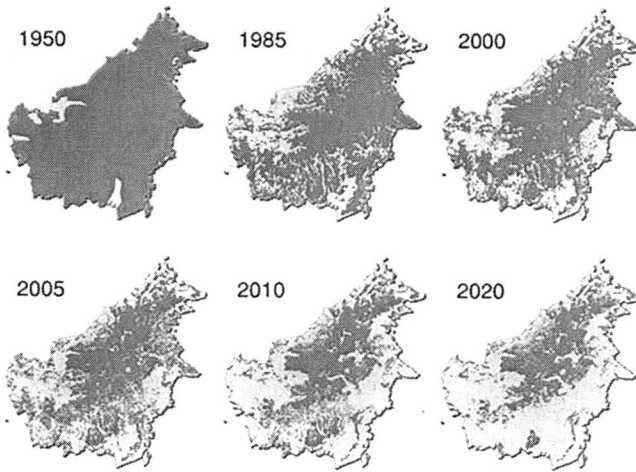

Figure 1.2 Changes in forest area in Borneo.

Source: UNEP/GRID-Arendal (2007).

http://maps.grida.no/go/graphic/extent-of-deforestation-in-borneo-1950-2005-and-projection-towards-2020

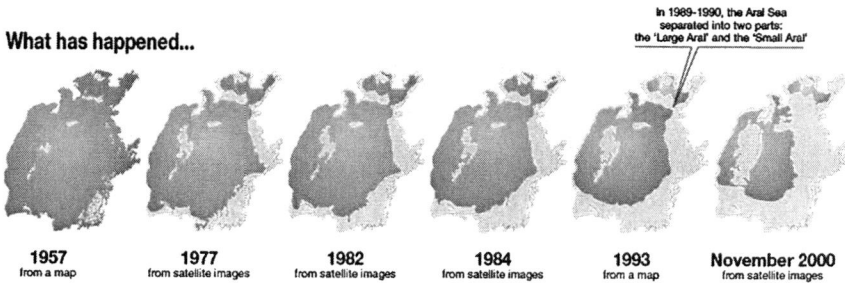

Figure 1.3 Changes in surface water area in the Aral Sea.

Source: UNEP/GRID-Arendal (2009).

http://maps.grida.no/go/graphic/the-disappearance-of-the-aral-sea

and dam construction. Climate change will have massive impacts on global fauna and flora due to changes in temperature and precipitation, and glacial retreat in the Himalayas and the Antarctic continent may devastate mountain systems and polar systems. Transportation of people and goods help invasive alien species to expand both intentionally and unintentionally, which may cause catastrophic damages to native food webs and vegetation, e.g. introduction of edible Nile perch (*Lates niloticus*) into Lake Victoria in East Africa in the 1950s has wiped out 200 species of native cichlid (*Cichlidae*) (Goldschmidt *et al.* 1993). Numerous cases of a decrease in species population due to overexploitation can be observed; American bison (*Bison bison*) has a drastic loss of population from tens of millions to less than 1,000 in the nineteenth century, and billions of passenger pigeons (*Ectopistes migratorius*) completely disappeared from the American continent in the early twentieth century. Pollution from chemicals and heavy metals has intolerable impacts on inland water systems, e.g. the spill incident of toxic materials from the mines in Spain in 1998 affected 4,600 hectares of land and water bodies, including Doñana National Park, and it exterminated an enormous number of fish equivalent to 30 tonnes in total (Bartolome and Vega 2002).

Additionally, MA (2005a) points out five indirect factors behind these direct drivers, i.e. population change, changes in economic activities, socio-political factors, cultural factors, and technological change. Increases in population and income will expand absolute resource use, and socio-political factors may have impacts on decision-making and education for ecosystem management. Recognition of ecosystems and consumption behaviors may have strong links to cultural and religious factors, and technological development can be one of the main drivers of an expansion of cultivated lands and overexploitation. Seeing these indirect factors from different viewpoints, market failures and policy failures are the two main indirect causes. The market failure in this context is to regard ecosystem services, excluding provisioning services, as having zero value, due to their nature of being public goods and services, resulting in overexploitation of natural stocks for benefit maximization and the underestimation of damage costs from pollution. In line with this viewpoint, the policy failures are persistence in granting subsidies, and trade structures which promote ecosystem degradation, and the absence of powerful international treaties for ecosystem conservation.

Unsustainable use of ecosystem services

Currently, 15 ecosystem services out of the 24 which were assessed in the MA currently show downward trends (see Table 1.2) (MA 2005a). With rapid population growth, improved productivity based on technical development and cropland expansion mean an increase in the production of crops, livestock, and aquaculture, while enhanced fishing capability and excess water intake for irrigation cause degradation of fish stocks and a decrease in freshwater resources, respectively. One-quarter of important commercial fish stock is currently overexploited, and one-quarter of water use in the world is supported by artificial water transfer and excess groundwater withdrawal (MA 2005a).

A decrease in forest area will reduce the climate regulation function, and the logging of riverine forests weakens regulation of soil erosion. Wetland ecosystem degradation reduces regulating services regarding water purification and waste assimilation, and degradation of mangroves and coral reefs scales down natural-hazard regulation and scales up the risks of damages from storms along coastlines. Current promotion of ecotourism increases recreational services on the one hand, but on the other hand intensive use of specific sites may degrade local ecosystems and reduce other ecosystem services there.

A temporary increase in provisioning services through enhancement of food production, overhunting, and exploitative logging will have positive effects on human well-being in terms of basic materials for a good life. On the contrary, reduced water purification will have negative impacts on health and the enlargement of damages from natural disasters will endanger security. Briefly, ecosystem services have trade-offs with each other; the negative impacts of increases in provisioning services on other ecosystem services should be subtracted from the benefits of direct use. Note that these trade-offs do not necessarily emerge within the same region and for the same person. For instance, development of shrimp farming is associated with mangroves being cut down in South East Asian countries; the benefits of provisioning services will be distributed to developed countries, while the costs of reduced fish stocks and increases in damages from natural disasters due to the mangroves degradation will be passed on to local people. In these cases, compensation from consumers in developed countries to producers in developing countries should be made, although these costs are not currently reflected in the product prices.

Degradation of ecosystem services squeezes the poor in the world. Degraded water purification services due to ecosystem destruction and pollution may increase the possibility of diseases from polluted drinking water and pathogen growth, thereby killing 1.8 million people per annum, a situation made worse with a scarcity of public sanitation (WHO 2007). Reduced fish stocks diminish cheap protein sources in developing countries and cause a drop in income and an increased risk of unemployment. Moreover, privatization of fishing rights, which used to be common among local people, led to deprivation of the rights of the weak who are dependent on natural resources (Sekhar 2004). Growing vulnerability to natural disasters due to ecosystem degradation is pointed out as one of the causes of an increased necessity for international assistance for natural disasters such as flooding, drought, earthquakes, and tsunamis (MA 2005a). Degradation of ecosystem services is an extremely important issue for the poor who are highly dependent on ecosystem services on the one hand, and conversely, poverty may be a great factor for increasing the load on the environment. Hence, policies will be required to break out of the vicious circle of poverty and to ease ecosystem degradation.

Ecosystem degradation also has negative impacts in developed countries. Extinction of plant species due to deforestation and development will mean the loss of opportunities to discover genetic resources for new drug developments; and a decrease in fish stock due to overexploitation may lower food production and employment potential in developed countries as well. Disappearance of forests and wetlands will increase risks of climate change, such as the rise in the sea level in coastal areas, higher intensity of natural disasters, and changes in fauna, flora, and crops. The increase in disease in developing countries may spill out to developed countries, and degradation of ecosystems in the biodiverse tropical forests will devalue ecotourism at those sites.

As such, unsustainable use of ecosystem services has the possibility to place tremendous negative impacts on human well-being. Furthermore, as it is impossible to reanimate extinct species, restoration of the lost ecosystem is extremely difficult, so the likelihood of exploitative use followed by restoration seems hard to accept. All in all, ecosystem services need to be used in a way which does not degrade the ecosystems; and ecosystem conservation that balances use and protection as well as sustainable use of ecosystem services will be necessary.

Policies for sustainable use of ecosystem services

Economic approach

As sustainable use of ecosystem services cannot be achieved in a day, various policies and activities will be required. MA (2005a) classifies these into six categories, namely, institutions and governance, economics and incentives, social and behavioral responses, technological responses, knowledge and cognitive responses, and the design of an effective decision-making process. Generally, these will be separated into two approaches here, i.e. economic and social approaches.

Economic activities such as land development, resource extraction, and international trade will degrade ecosystem services directly, and support for unsustainable economic activities including taxes and subsidies will be indirect factors for ecosystem degradation. In order to halt the unsustainable use of ecosystem services, it will be necessary to build economic structures in which sustainable use of ecosystem services will provide greater economic efficiency than unsustainable use thereof. In doing so, changes in economic systems, including the creation of incentives and the elimination of the subsidies encouraging ecosystem degradation, will be needed.

Certification schemes and green procurement exemplify activities to reduce the negative impacts of consumption and production on ecosystem services. The former is the so-called eco-label, represented by the Forest Stewardship Council (FSC) certification and Marine Stewardship Council (MSC) certification, which provide certification marks for the outputs of forestry and fisheries produced in a sustainable manner. These schemes are thought to provide producers with incentives for shifting to sustainable production methods, because a price premium similar to branded products in a common market can be expected. Although the pricing effect is a controversial issue, Wakamatsu *et al.* (2010) demonstrate the Japanese people's willingness to pay the price premium of the MSC products, and Managi *et al.* (2008) show the consumer demands for eco-friendly agricultural products.

Green procurement requires an organization to procure eco-products in a certain proportion to specific purchased goods, aiming to reduce environmental impacts through creating demand for eco-products. For instance, insisting on the procurement of recycled paper by compounding old paper will increase market demand for recycled paper and may reduce the rate of deforestation. Although some may insist that recycled paper made solely from old paper has the possibility of increasing environmental impacts, it can be diminished by accepting a mixture with FSC products. Green procurement is currently an obligation only for public organizations, but the market size of eco-products is expected to expand in the future, as a growing number of private sectors implement similar activities.

Fundamentally, the internalization of the economic value of ecosystem services currently regarded as externalities (i.e. pricing them in the market) will be required to correct market failures. As stated above, this is because insufficient recognition of the economic value of ecosystem services prioritizes short-term benefits from development, resulting in a reduction of long-term benefits from the natural environment. Economic valuation of ecosystem services makes the benefits visible in a simple manner. Also, this will allow comparison between the benefits from ecosystem services and profits from development and exploitation, thereby creating economic incentives for ecosystem conservation in the case of the former exceeding the latter. In order to realize the internalization of the economic value of ecosystem services, it will be necessary to provide a correct evaluation thereof, and to formulate legal and institutional policies including payment for ecosystem services providers and burden-sharing of the costs of ecosystem degradation.

Social approach

A social approach to the sustainable use of ecosystem services covers a wide range of issues from enhancement of governance to improvement in communication and multi-stakeholder participation. Here, four issues will be introduced, namely, conclusion of multi-lateral environmental treaties, designation of protected areas, community involvement, and utilization of traditional knowledge.

The CBD is the most comprehensive multilateral environmental treaty regarding biodiversity and Article 1 clearly mentions "the conservation of biological diversity, the sustainable use of its components and the fair and equitable sharing of the benefits arising out of the utilization of genetic resources" as its objective (UN 1992). Although it has some problems, such as non-ratification by the United States of America, successful promotion of the importance of sustainable use can be evaluated. The concrete measures for assurance of their effectiveness and the realization of sustainable use of ecosystem services will need to be considered further.

Since the establishment of the target at the World Parks Congress in Bali, Indonesia, in 1982 to designate more than ten percent of terrestrial areas of each country as protected areas, there has been an increase in both the numbers and the areas of them (see Figure 1.4) (IUCN and UNEP-WCMC 2009). In response to the problem that strict protection by designating protected areas may decrease the possibility for direct use of natural resources in the local sites, there have been recent attempts to strike a balance between the use and protection of ecosystems through separation of the area into central-core protected areas and surrounding usable buffer zones. Additionally, international cooperation on lowering the burden for local residents will be promoted, as regulating services provided by protected areas will benefit not only local residents but also overseas countries.

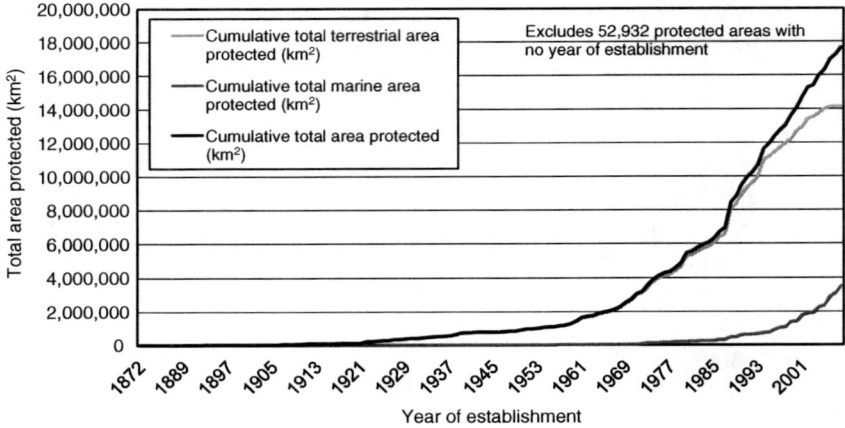

The chart shows a y-axis labeled "Total area protected (km²)" with values from 0 to 20,000,000 in increments of 2,000,000, and an x-axis labeled "Year of establishment" with years from 1872 to 2001. The legend indicates:
- Cumulative total terrestrial area protected (km²)
- Cumulative total marine area protected (km²)
- Cumulative total area protected (km²)

A note reads: "Excludes 52,932 protected areas with no year of establishment"

Figure 1.4 Growth in nationally designated protected areas (1872–2008).

In contrast to this top-down approach, a growing number of voluntary activities by communities for the purpose of sustainable use of ecosystem services have been observed. In the case of the CAMPFIRE program in Zimbabwe, the community decided to manage its land for game hunting and in the light of conflicts between residents and wildlife over agricultural crops, it implemented various actions including the installation of electric fences and the employment of rangers. As a result of these activities, positive effects could be obtained such as a rise in income, relaxation of conflict, an increase in the wildlife population, and a reduction in poaching. This is a good example which shows that communities can improve efficiency in natural resource management.

Traditional knowledge stored by indigenous people varies widely from the protection of specific wildlife, management of diverse species, implementation of resource circulation, establishment of a transitional agricultural system, transmission of knowledge through lore and rites, to restriction by taboos. Some of them are useful for sustainable use of ecosystem services, inter alia, monitoring of ecosystem status through cumulative daily activities, life arrangement and adaptation to environmental conditions, and adaptive and dynamic ecosystem management. Supplementing traditional knowledge with scientific expertise will help to bring about the sustainable use of ecosystem services.

Conclusion of Nagoya Protocol and the future of ecosystem services

The Nagoya Protocol and the Aichi Target were concluded at the CBD-COP10 on 30 October 2010. The former contributes to international rule-making on fair and equitable benefit-sharing from the use of genetic resources, mentioning

Table 1.3 Twenty goals in the Aichi Target by 2020

Strategic Goal A: Address the underlying causes of biodiversity loss by mainstreaming biodiversity across government and society.

Target 1   People are aware of the values of biodiversity and the steps they can take to conserve and use it sustainably.

Target 2   Biodiversity values have been integrated into national and local development and poverty-reduction strategies and planning processes and are being incorporated into national accounting, as appropriate, and reporting systems.

Target 3   Incentives, including subsidies, harmful to biodiversity are eliminated, phased out, or reformed in order to minimize or avoid negative impacts; and positive incentives for the conservation and sustainable use of biodiversity are developed and applied.

Target 4   Governments, business, and stakeholders at all levels have taken steps to achieve or have implemented plans for sustainable production and consumption and have kept the impacts of the use of natural resources well within safe ecological limits.

Strategic Goal B: Reduce the direct pressures on biodiversity and promote sustainable use.

Target 5   The rate of loss of all natural habitats, including forests, is at least halved and where feasible brought close to zero, and degradation and fragmentation are significantly reduced.

Target 6   All fish and invertebrate stocks and aquatic plants are managed and harvested sustainably, legally, and applying ecosystem-based approaches, so that overfishing is avoided, and recovery plans and measures are in place for all depleted species.

Target 7   Areas under agriculture, aquaculture, and forestry are managed sustainably, ensuring conservation of biodiversity.

Target 8   Pollution, including from excess nutrients, has been brought to levels that are not detrimental to ecosystem function and biodiversity.

Target 9   Invasive alien species and pathways are identified and prioritized, priority species are controlled or eradicated, and measures are in place to manage pathways to prevent their introduction and establishment.

Target 10   By 2015, the multiple anthropogenic pressures on coral reefs and other vulnerable ecosystems impacted by climate change or ocean acidification are minimized, so as to maintain their integrity and functioning.

Strategic Goal C: To improve the status of biodiversity by safeguarding ecosystems, species, and genetic diversity.

Target 11   At least 17% of terrestrial and inland water, and 10% of coastal and marine areas, especially areas of particular importance for biodiversity and ecosystem services, are conserved through being effectively and equitably managed.

Target 12   The extinction of known threatened species has been prevented and their conservation status has been improved and sustained.

Target 13   The genetic diversity of cultivated plants and farmed and domesticated animals and of their wild relatives is maintained, and strategies have been developed and implemented for minimizing genetic erosion and safeguarding their genetic diversity.

Continued

Table 1.3 Continued

Strategic Goal D: Enhance the benefits to all from biodiversity and ecosystem services.

Target 14	Ecosystems that provide essential services, including services related to water, and that contribute to health, livelihoods, and well-being are restored and safeguarded, taking into account the needs of women, indigenous and local communities, and the poor and vulnerable.
Target 15	Ecosystem resilience and the contribution of biodiversity to carbon stocks has been enhanced, through conservation and restoration, including restoration of at least 15% of degraded ecosystems, thereby contributing to climate-change mitigation and adaptation and to combating desertification.
Target 16	By 2015, the Nagoya Protocol on Access to Genetic Resources and the Fair and Equitable Sharing of Benefits Arising from their Utilization is in force and operational, consistent with national legislation.

Strategic Goal E: Enhance implementation through participatory planning, knowledge management, and capacity building.

Target 17	By 2015 each Party has developed, adopted as a policy instrument, and has commenced implementing an effective, participatory, and updated national biodiversity strategy and action plan.
Target 18	The traditional knowledge, innovations, and practices of indigenous and local communities relevant for the conservation and sustainable use of biodiversity, and their customary use of biological resources, are respected, subject to national legislation and relevant international obligations.
Target 19	Knowledge, the science base, and technologies relating to biodiversity, its values, functioning, status and trends, and the consequences of its loss, are improved, widely shared and transferred, and applied.
Target 20	The mobilization of financial resources for effectively implementing the Strategic Plan for Biodiversity 2011–2020 from all sources, and in accordance with the consolidated and agreed process in the Strategy for Resource Mobilization, should increase substantially from the current levels.

Source: CBD website: Aichi Biodiversity Target, revised by the authors http://www.cbd.int/sp/targets/IUCN and UNEP-WMC (2009).

prior informed consent of the countries providing resources, and implementation of legal measures in each country. The latter is the newly concluded goal for ecosystem conservation from the year 2011 in response to the failure of the 2010 Biodiversity Target, which aims to achieve "by 2010 a significant reduction of the current rate of biodiversity loss at the global, regional and national level" (CBD website). The Aichi Target sets 20 individual goals for each country to achieve, including the designation of 17 percent of terrestrial areas and ten percent of marine areas as protected areas (see Table 1.3). Additionally, several other decisions were made such as the creation of the Intergovernmental Science and Policy Platform on Biodiversity and Ecosystem Services (IPBES), establishment of the strategy for world plant conservation, and recognition of the importance of rice fields.

As shown above, global biodiversity conservation is steadily progressing step by step. Further acceleration along this path is expected through effective policies, more participation of private sectors and civil society, and further research on biodiversity and ecosystem services.

References

Bartolome, J. and Vega, I., 2002, *Mining in Donana: Learned Lessons*, Madrid, WWF Spain.

Ehrlich, P. R. and Ehrlich, A. H., 1981, *Extinction: The Causes and Consequences of the Disappearance of Species*, New York, Random House.

Convention of Biological Diversity (CBD), Aichi Biodiversity Targets, available at http://www.cbd.int/sp/targets/ (Accessed: 6 February 2012).

Elmqvist, T., Folke, C., Nystrom, M., Peterson, G., Bengtsson, J., Walker, B. and Norberg, J., 2003, "Response diversity, ecosystem change, and resilience," *Frontiers in Ecology and the Environment*, 1(9): 488–494.

Goldschmidt, T., Witte, F. and Wanink, J., 1993, "Cascading effects of the introduced Nile Perch on the detritivorous/phytoplanktivorous species in the sublittoral areas of Lake Victoria," *Conservation Biology*, 7(3): 686–700.

IUCN and UNEP-WCMC, 2009, The World Database on Protected Areas (WDPA), January 2009.

Managi, S., 2011, "Productivity measures and effects from subsidies and trade: An empirical analysis for Japan's forestry," *Applied Economics*, 42(30): 3871–3883.

Managi, S. Yamamoto, Y., Iwamoto, H. and Masuda, K., 2008, "Valuing the influence of underlying attitudes and the demand for organic milk in Japan," *Agricultural Economics*, 39(3): 339–348.

Millennium Ecosystem Assessment (MA), 2005a, *Ecosystems and Human Well-Being, Synthesis*, Washington, D.C.: Island Press.

——, 2005b, *Ecosystems and Human Well-Being, Volume 1: Current State and Trend*, Washington, D.C., Island Press.

Peterson, G., Allen, C. R. and Holling, C. S., 1998, "Ecological resilience, biodiversity, and scale," *Ecosystems*, 1(1): 6–18.

Pidwirny M., 1999, "Geography 210: Introduction to environmental issues," cited at the *Vital Water Graphics*, prepared by UNEP. http://www.unep.org/dewa/vitalwater/article115.html (Accessed: 10 December 2010).

Sekhar, N. U., 2004, "Fisheries in Chilika lake: How community access and control impacts their management," *Journal of Environmental Management*, 73(3): 257–266.

Tilman D., Polasky, S. and Lehman, C., 2005, "Diversity, productivity and temporal stability in the economies of humans and nature," *Journal of Environmental Economics and Management*, 49: 405–426.

United Nations (UN), 1992, Convention of Biological Diversity, available at http://www.cbd.int/doc/legal/cbd-en.pdf (Accessed: 6 February 2012).

UNEP/GRID-Arendal, 2007, *Extent of deforestation in Borneo 1950–2005, and projection towards 2020*, UNEP/GRID-Arendal Maps and Graphics Library. http://maps.grida.no/go/graphic/ extent-of-deforestation- in-borneo-1950-2005- and-projection-towards-2020 (Accessed: 2 December 2010).

UNEP/GRID-Arendal, 2009, *The disappearance of the Aral Sea*, UNEP/GRID-Arendal Maps and Graphics Library. http://maps.grida.no/go/graphic/the-disappearance-of-the-aral-sea (Accessed: 22 April 2011).

UNEP/WCMC, 2009, *World Database on Protected Areas (WDPA): January 2009*, Cambridge, UK: UNEP-WCMC. http://www.wdpa.org/resources/statistics/2009MDG_Growth_Chart.gif (Accessed: 21 April 2011).

Wakamatsu, H., Uchida, H., Roheim, C. A., Anderson C. M. and Managi, S., 2010, *Consumer Preferences for Ecolabeled Seafood in Japan: Influence of Information on Preferences*, presented at the XVth biennial conference of the International Institute of Fisheries Economics and Trade, Montpelier, France, July, 2010.

WHO, 2007, *Combating waterborne disease at the household level*. http://www.who.int/water_sanitation_health/publications/combating_diseasepart1lowres.pdf (Accessed: 10 December 2010).

2 Convention on Biological Diversity and other initiatives for worldwide protection of biological diversity and ecosystem services

Tania Ray Bhattacharya and
Shunsuke Managi

General overview of the Convention on Biological Diversity

Earth's enriched biological resources are of tremendous value to present and future generations of human beings. But unfortunately, due to various environmental reasons impacted on by human activities and otherwise, ecosystems and species-rich biological diversity are presently under great threat and species extinction continues at an alarming rate. In response, various measures are adopted to protect the biodiversity level. The clearest antecedent is the case of "botanizing" found in the ages of Empire and explorations when the collectors used to collect rare species from overseas. The concepts introduced in the field of science and technology studies (STS) during this time can still be used to examine the complex field of biodiversity (Escobar 1998). Table 2.1 shows the varieties of interactions of different drivers of an ecosystem and their corresponding impacts on the ecosystem.

"Biological diversity is the key to the maintenance of the world as we know it" (Wilson 1993). The United Nations Environment Program (UNEP) established the *Ad Hoc* Working Group of Experts on Biological Diversity in November 1988 to explore the requirement of introducing an international convention on biological diversity. An *Ad Hoc* Working Group of Technical and Legal Experts was formed in May 1989 to set up the legal instruments for the protection and sustainable use of biological diversity. This group decided to take care of the cost and benefit sharing between the developed and developing countries, and also to support the innovations of the local people. From February 1991 onwards this group was known as the Intergovernmental Negotiating Committee (Convention on Biological Diversity 2011a). The textual origin of the Convention of Biological Diversity (CBD) was dated 22 May 1992 at the Nairobi conference. More than 100 countries and many international organizations participated in the conference. The CBD was signed at the Earth Summit in Rio de Janeiro on 5 June 1992. Sixteen countries signed the CBD on that day. And other 141 countries signed the CBD in Brazil on 14 June 1992. Now there are 193 Parties in the CBD, including

Table 2.1 Interaction of ecosystem drivers and corresponding impacts

Type of region	Climate type	Habitat change	Climate change	Invasive species	Over exploitation	Pollution (nitrogen, phosphorus)
Forest	Boreal	↗	↑	↗	→	↑
	Temperate	↘	↑	↑	→	↑
	Tropical	↑	↑	↑	↗	↑
Dry land	Temperate grassland	↗	↑	→	→	↑
	Mediterranean	↗	↑	↑	→	↑
	Tropical grassland and savanna	↗	↑	↑	→	↑
	Desert	→	↑	→	→	↑
Inland water		↑	↑	↑	→	↑
Coastal		↗	↑	↗	↗	↑
Marine		↑	↑	→	↗	↑
Island		→	↑	→	→	↑
Mountain		→	↑	→	→	↑
Polar		↗	↑	→	↗	↑

Source: Millennium Ecosystem Assessment, (2005).

Note: Table 2.1 indicates how different drivers for ecosystems and biodiversity affect different types of ecosystems in the world. Direction of the arrows indicates the direction of impacts per se.

168 Signatories. The United States of America has signed but has not yet ratified the CBD.

The CBD in general provides a worldwide legal platform for action on biodiversity. The Parties of the Convention's governing body called the Conference of the Parties (COP) meet every two years, or as required, for adopting different programs of work, monitoring progress in the implementation of the Convention and achieving its objectives and, finally, to provide policy guidance. Another important organ of this process is the Subsidiary Body on Scientific, Technical, and Technological Advice (SBSTTA). It is made up of government representatives from relevant fields, observers from non-Party governments, the scientific community, and other relevant organizations that provide recommendations to the COP on the technical aspects of the implementation of the convention. SBSTTA works as the functioning wing of the COP to implement the decisions taken in the COP meeting. There are ad hoc open-ended Working Groups too dealing with specific issues with a limited mandate and period of time. Under the convention there are four major working groups working primarily in four different major areas of concern as follows.

- The Working Group on Access and Benefit-Sharing (ABS) is currently the forum for negotiating an international regime on access and benefit sharing.
- The Working Group on Article 8(j) deals with issues related to the protection of traditional knowledge.

- The Working Group on Protected Areas is guiding and monitoring the implementation of the program of work on protected areas.
- The Working Group on the Review of the Implementation of the Convention (WGRI) examines the implementation of the convention, including national biodiversity strategies and action plans.

Working groups make recommendations to the COP, and, as is the case for the Working Group on Access and Benefit-Sharing, may also provide a forum for negotiations on a particular instrument under the convention. The COP and SBSTTA may also establish expert groups or call for the organization by the secretariat of liaison groups, workshops, and other meetings. Participants in these meetings are usually experts nominated by governments, as well as representatives of international organizations, local and indigenous communities, and other bodies. Unlike SBSTTA and the open-ended Working Groups these are usually not considered as intergovernmental meetings. The purpose of these meetings varies: expert groups may provide scientific assessments, for example, while workshops may be used for training or capacity building. Liaison groups advise the secretariat or act in cooperation with other conventions and organizations.

Access and benefit sharing, intellectual property rights, indigenous knowledge, and biopiracy

The third objective of the CBD is access and benefit sharing of ecosystem services which has not yet been achieved to the desired level, and the degree of its attainment is below the desired level (Medaglia 2010). It has been observed that global biodiversity is greater near the equator and less near the poles, which apparently creates a huge difference between its use and conservation (G/Egziabher 2000). Biodiversity level is highly concentrated in the areas of the developing countries rather than the developed countries, and the developing world is more dependent on local biodiversity resources for their basic needs. Two-thirds of Earth's plant species including 35,000 medicinal plants are from the developing countries (Crucible Group 1994). The world's variety of food production depends essentially on biological diversity including various animal and plant species, savannas and forest, and marine ecosystems of the developing countries. Figure 2.1 shows the geographical spread of biodiversity hotspots.

Indigenous knowledge (IK) is the local knowledge which is unique to a given culture or society; it is the information base for a society, which facilitates communication and decision-making. The industrialized developed countries have advanced technologies to make commercialized products by manipulating the rich biological resources from the developing countries. The commercialization of naturally occurring biological materials like plant substances or genetic cell lines by a technologically advanced country or by an organization without proper compensation to the nations in whose territory the materials were originally discovered, is known as "biopiracy."

Figure 2.1 Global spread of biodiversity hotspots.

The developed countries have their own argument that these resources are common properties for human beings. But contrarily, their production from these resources from the developing countries are defined as "private heritage" which are protected by mechanisms called patents which prevent other countries from deriving benefits from these products. These patents are referred to as Intellectual Property Rights (IPR). The justification of IPR and Indigenous Knowledge System (IKS) is a debatable issue between the developed and the developing countries.

Aside from the distribution of biological diversity and ecosystem services among nations, importance should also be given to distribution at the regional and national levels. The design of benefit-distribution mechanisms should incorporate this idea. Some communities might migrate from biodiversity rich areas to marginal areas, despite their traditional contribution to the conservation of the ecosystems and the level of biodiversity.

Trade-related aspects of intellectual property rights (TRIPS) agreements

TRIPS is one of the most important multilateral agreements on intellectual property. It has a most significant impact on the pharmaceutical sector and medicines. The TRIPS Agreement was signed in 1995 and introduced global minimum standards for protection and enforcement of almost all forms of IPR which include patents. In this agreement, over 40 countries did not agree on the issue of patent protection for pharmaceutical products. The TRIPS Agreement now made it mandatory that all WTO members should adopt laws which support minimum standards of IPR protection, along with obligations for the enforcement of IPR. However, TRIPS also allows countries to have their freedom to modify their regulations and it also provides various incentives to the countries for formulating their national legislation to ensure balance between the providing the incentives for future inventions of new drugs and affordable accessibility to existing medicines.

The global valuation of biodiversity which includes IPR over the biological forms adds more meaning to the reasoning behind sustainable development. IPR could be considered as the fourth pillar that the modern world is adding to Adam Smith's trilogy of labor, capital, and natural resources. IPR over life drive the privatization of biodiversity and agricultural resources. The Agreement on Trade-related Aspects of IPR (the TRIPS Agreement), being one of the leading international documents in the field, states in its preamble that, "intellectual property rights are private rights." In fact, the whole point of patenting, which is a mainstream form of IPR, is to exclude others from access to information resources. Therefore, IPR enforce the privatization of resources, research, knowledge, and technology on a wide scale. By supporting these trends, the global green development is actually proposing to privatize and sell nature to preserve it. Table 2.2 shows the global valuation of ecosystems and biodiversity.

Cartagena Protocol

The Cartagena Protocol on biosafety of the CBD is an international treaty which controls the movement of Living Modified Organisms (LMOs) resulting from modern biotechnology from one country to another. The protocol also establishes a Biosafety Clearing-House to facilitate the exchange of information on LMO and to assist countries in implementing the protocol. This protocol was adopted on 29 January 2000 as a supplementary agreement to the CBD and was brought into force on 11 September 2003 (Convention on Biological Diversity 2011b).

Financial mechanisms under the CBD

The major stumbling block for the successful implementation of the CBD is the financing of the conservation process and its sustainable implementation. Globally, there are certain major initiatives in the context of financing the implementation strategies to achieve the targets. The following sub sections discuss such major initiatives in the world.

The Global Environment Facility and the CBD

The Global Environment Facility (GEF) was established in 1994 by the United Nations Development Program (UNDP), UNEP, and the World Bank to promote international cooperation and to finance actions to deal with four critical threats to the global environment: biodiversity loss, climate change, degradation of international waters, and ozone depletion. The GEF funding program mainly deals with the biodiversity-related issues in the dry land areas, coastal, marine, and inland freshwater systems, mountains and different types of forest ecosystems, such as cloud forests, tropical rainforests, dry forests, temperate forests, boreal forests, and mangrove forests. However, the GEF implements the guidance of the CBD through the following main activities.

Table 2.2 Global valuation of ecosystem and biodiversity

Global template	Area (Mill. km²)	Mean ESV (USD/km²/yr)	Total ESV (billion US$/year)				Percentage above random	Concordance index (%)
			Observed	Random	Max.			
High-biodiversity wilderness areas (Mittermeier et al. 2003)	11.5	200,720	2,314	701	4,708	230	40.3
Frontier forests (Bryant et al. 1997)	13.2	188,224	2487	803	5,151	210	38.7
Most proactive	7.6	217,356	1,659	464	3,681	257	37.1
Global 200 eco-regions (Olson and Dinnerstein 1998)	53.8	86,857	4,671	3,270	7,466	43	33.4
Last of the wild (Sanderson et al. 1997)	35.0	98,356	3,440	2,127	6,838	62	27.9
Megadiversity countries (Mittermeier et al. 1997)	49.8	77,457	3,860	3,031	7,340	27	19.2
Endemic bird areas (Stattersfield et al. 1998)	13.8	88,710	1,222	838	5,301	46	8.6
Centers of plant diversity (WWF and IUCN 1994–1997)	12.2	83,779	1,023	743	4,888	38	6.8
Most reactive	12.1	76,057	917	734	4,849	25	4.5
Biodiversity hotspots (Mittermeier et al. 2004)	23.0	69,071	1,588	1,398	6,289	14	3.9
Random terrestrial km²	NA	60,813	NA	NA	NA	NA	0
Crisis eco-regions (Hoekstra et al. 2005)	42.7	46,038	1,967	2,598	7,112	−24	−14.0

Source: Turner et al. (2007).

- Operational programs for the long-term protection of biodiversity, the sustainable use of biological resources, and the sharing of benefits that arise from the use of genetic resources
- Enabling activities that assist countries in producing their National Biodiversity Strategies, Action Plans, and National Reports
- Short-term response measures that offer cost-effective opportunities to conserve and sustainably manage biodiversity and its resources.

Apart from the above-mentioned different programs, notable areas of GEF funding, conservation of the natural world heritage sites (designated by UNESCO), nature conservation with improving people's standard of living, as in Mexico and the countries of Central America, are helping to restore the Meso-American Biological Corridor. Moreover, the GEF funding also works for the humid tropics, Brazilian Amazon, and Guyana Shield, as well as for countries such as Cambodia, Indonesia, and the Lao People's Democratic Republic where conservation of tropical rain forests are promoted.

Multilateral initiatives in protecting biodiversity and ecosystem services

The conservation and protection activities of biological diversity are not only a complex matter which crosses over the national and geographical territories and generations together, but they are also very expensive to perform. As the biodiversity hotspots are mainly in the tropical regions of the Earth which are either developing or least developed in nature, financing is a major issue for them to have a systematic approach to conserve ecosystems and biodiversity. Ironically, the indigenous people whose livelihood primarily depends upon the quality, quantity, and continuity of the critical level of biodiversity are either not aware of the importance of the native species around them, or they do not have sufficient funds to conserve them in a sustainable manner. Moreover, the deterioration of critical species is often caused by industrialization which has no direct link to the local people's livelihood. As a matter of fact, the conservation of biological diversity has become an issue of multilateral negotiation. In this context, we have discussed various initiatives taken by the major multilateral organizations like the World Bank, the Asian Development Bank, and the African Development Bank, etc. who are actively involved in the process of the implementation of CDB using various standard and innovative financing mechanisms.

The World Bank initiatives

The World Bank is one of the major multilateral funding organizations in the world actively involved in the process of international initiatives for the conservation of biodiversity and ecosystems services. The World Bank's initiative towards biodiversity funding peaked nearly 20 years ago, when the GEF, designed to promote the recognition of global public goods, was being implemented. The GEF has been the central source of support grants implemented by the World Bank, and has leveraged co-financing from The International Bank for Reconstruction and Development (IBRD) and International Development Association (IDA) loans as well as other co-financiers. However, presently the World Bank Group (WBG) has become the largest biodiversity funder in the world. The bank has worked directly with 122 developing countries, as well as through a range of regional and global partnerships to protect and manage biodiversity. Between 1988 and 2009, the bank provided around 2 billion US dollars as loans, and over 1.4 US billion dollars in GEF resources, and it leveraged 2.9 billion US dollars in co-financing resulting in a total investment portfolio of more than 6.3 US billion dollars. The WBG portfolio mainly supports local activities aimed at the protection of small but critical habitats with communities and indigenous people establishing and managing national protected areas and the trust funds dedicated to these protection activities. The portfolio is increasingly focusing on mainstreaming biodiversity into forestry, coastal zone management, and agriculture, and also the improvement of natural resource management. The World Bank also supports regional and global initiatives for the stakeholders' capacity-building, and enhancement of their awareness to join activities for natural resources conservation. The client countries of the World Bank have doubled the total terrestrial protected areas within the last 20 years which now cover up to about 14 percent of the Earth's land surface. Some of the major initiatives of the World Bank's Global and Regional Partnerships to Save Endangered Species and Ecosystems are as follows.

- The Global Tiger Initiative (GTI).
- The Critical Ecosystems Partnership Fund with Conservation International (CI).
- The Amazon Region Protected Areas Project in Brazil.
- The Save our Species Program.
- The International Consortium on Combating Wildlife Crime (ICCWC).

The Asian Development Bank initiatives

In 1989, the Asian Development Bank (ADB) management issued a directive meant to integrate biodiversity conservation into its different project designs. This directive also highlighted the importance of biodiversity conservation in achieving sustainable economic development. As a major provider of financial and advisory services for environmental projects and programs in Asia and the Pacific, the ADB assists its developing member countries to cope with their obligations to the CBD and to other international agreements. Two major features of the CBD are of major relevance to development institutions like the ADB. The first is to identify biodiversity conservation as an important measure for sustaining the social and economic development of countries. The second is the importance given to sound policies (e.g. pricing, taxation, land tenure) and the effective institutional and social arrangements (e.g. laws, regulations, and the roles of the state, private sector, NGOs, local communities, and indigenous people) needed for achieving effective biodiversity conservation. Until the year 2000, the ADB invested approximately 1.3 billion US dollars for 35 biodiversity conservation projects in nine of its developing member countries.

The African Development Bank initiatives for biodiversity and ecosystem conservation

The African Development Bank (AfDB) believes that green growth in Africa would depend on sustained commitment to the maintenance of the existing level of biodiversity and ecosystems, as well as the various services they provide to local communities. The bank addresses biodiversity and ecosystem protection services as major inputs to its sustainable development policies. The Integrated Water Resources Management policy of the AfDB states that development in agriculture, food production, animal husbandry and related biotechnology, forestry, and medicines will depend on the preservation of biodiversity. The 2004 Environmental Policy of the bank addresses biodiversity as an essential public good. The bank is also involved in the implementation of different activities such as the Congo Basin Forest Fund, the African Water Facility, Central Africa REDD and REDD+ projects, etc. Reported funding and investments in various forest projects supported by the AfDB along with other major ODA donors, including the OECD countries, the World Bank, IADB, and GEF are estimated as being around 1.1 billion US dollars in 2004. The AfDB also invested around 50 million pounds in collaboration with the British government for the conservation of the Congo Basin Forest project in 2008, which is one of the largest biodiversity and ecosystem conservation projects in Africa. Further, the AfDB was involved in projects such as the Selous Game Reserve Management project in Tanzania and in the regional projects for sustainable management of endemic ruminant livestock in West Africa.

Country perspectives on biodiversity and ecosystem conservation

Depending upon the social, environmental, and economic conditions of the countries, the world has different levels of perception, action, and planning for CBD. It is quite obvious that the conservation of biodiversity will always take a lower priority compared to the need for basic amenities for the survival of the human race. This has been reflected in the international negotiation processes over the last ten COP meetings. It is hard to believe that there is any real progress in the area of convergence of ideas and thinking of the parties, but the reality is that much less progress has been made, so far, in terms of achieving a multiregional agreement on the issues of conservation. Nevertheless, India, Indonesia, China, Japan, the United States, and European countries are the major stakeholders and are parties in the CBD processes. The following sections briefly describe the national issues for selected countries.

Indonesia

Indonesia is one of the 17 "mega diverse" countries, which consist of two of the world's 25 "hotspots," 18 World Wildlife Fund's "Global 200" eco regions, and

24 of Bird Life International's "Endemic Bird Areas." Also, there are ten percent of the world's flowering species in Indonesia and this country is recognized as one of the world's centers for agro-biodiversity of plant cultivars and domesticated livestock. Indonesia is also recognized worldwide as a major center of species diversity of hard corals and many groups of reef-associated flora and fauna. However, Indonesia's enriched biodiversity level is under threat from rapid landscape change, pollution, and over-harvesting and is rapidly becoming degraded.

Although Indonesia has not designed any specific country-oriented targets, it has several programs, strategies, and action plans for biodiversity protection for some of the relevant sectors which could be used in accordance with the conservation of biodiversity. Different plans include: expanding Marine Protected Areas (MPA) by approximately five million hectares by 2010; capacity-building programs for better management of the MPAs; and establishing global networks and partnerships for trans-boundary MPAs. Policies and actions related to forest protection include: effective management of the national parks; restoring populations of endangered, threatened, and critical species in the conservation areas; rehabilitating or reintroducing flagship species; and strategic development plans for maintaining genetic diversity and species purity. Forest rehabilitation enhancement has been going on effectively in the entire country. Community-based forestry projects covering two million hectares have also been established. Research and development of the genetic resource potential of plants and critical and endangered animals, establishment of a gene bank of plant and animal genetic resources, and application of traditional knowledge in the utilization of biological resources, etc., are included in the programs of the Ministry of Agriculture.

India

India is also one of the 17 "mega diverse" countries and is enriched with different ecosystems such as forests, grasslands, wetlands, coastal, marine, and desert areas. Major causes of biodiversity losses in India include land conversion for agricultural uses, unplanned development, opening of roads, overgrazing, fire, pollution, the introduction and spread of exotic diseases, excessive siltation, dredging, and reclamation of water bodies, mining, and industrialization. Presently, the species such as the Indian cheetah, Indian rhino, pink-headed duck, forest owlet, and the Himalayan mountain quail are reported to have become extinct and several other species are identified as vulnerable or endangered in India. India's National Policy and Macro-level Action Strategy on Biodiversity 1999 aims at the conservation and sustainable use of biological diversity, which includes regeneration and the rehabilitation of threatened species, participation of state governments, communities, people, NGOs, industry and other stakeholders, and research and development to understand the value of biodiversity. India is rich in traditional knowledge associated with biological resources. There is coded traditional knowledge, as in the texts of Indian systems of medicine, and also non-coded knowledge, which is mainly oral and undocumented. Recently, a Traditional Knowledge Digital Library (TKDL) was developed in India, which is a computerized database of documented information on Indian medicine systems. The main objectives of the TKDL are: (i) preservation of traditional knowledge; (ii) prevention of misappropriation of traditional knowledge; and (iii) creation of linkages with modern science for initiating active research for the invention of new drugs. Fair and Equitable Benefit Sharing in respect of biological material and traditional knowledge with the usage of such biological material is already proposed in the National Environment Policy to enable the country and the local communities, respectively, to derive economic benefits from providing access. Further, India has taken three significant legislative measures related to access and benefit sharing. India has enacted the Biological Diversity Act, 2002, which primarily aims at regulating access to biological resources and associated traditional knowledge to ensure equitable sharing of benefits from this usage, which is in accordance with the provision of Article 15 of the CBD. The Plant Varieties Protection and Farmers' Rights Act PVPFRA, 2001, and the PVPFR Rules, 2003, which are implemented for the protection of plant breeders' rights over the new varieties developed by them and the entitlement of farmers to register and to save, breed, use, exchange, share, or sell the plant varieties developed, improved, and maintained by them over many generations. There are also The Patent Second Amendment Act, 2002, and the Patent Third Amendment Act, 2005, for the exclusion of plants and animals from the purview of patentability (Section 4e); and the exclusion of an invention which in effect is traditional knowledge from patentability (Section 4p).

China

China being a "mega diverse" country possesses almost all types of ecosystems which include agricultural, forest, inland water, marine and coastal, dryland and semi-arid, mountain, and island. China is home to approximately 6,347 species of vertebrates, including 581 animal, 1,244 bird, 376 reptile, 284 amphibian, and 3,862 fish species. The major threats to biodiversity and ecosystems in China include over usage of lands, overconsumption of wild animal and plant species; destruction of habitats; lack of protection of some wild animals and plants; invasive alien species; deforestation and pollution.

China's national biodiversity strategy action plan aims to undertake effective measures to curb further damage to the natural environment and resources in China and to mitigate this serious situation. The action plan comprises seven major objectives aimed at: strengthening the fundamental studies of biodiversity; protecting wild species of biodiversity importance; improving the networks and management of protected areas; protection of the genetic resources of crops and domesticated animals; establishing national networks of biodiversity monitoring and information; and bridging the concept of biodiversity conservation; and sustainable development. Different priority programs and conservation activities include: assessment of the status of biodiversity and its economic values; assessment of the representativeness and effectiveness of protected areas and identifying needs for establishing new protected areas; identifying priority wild animals for protection based on their biodiversity importance and the level of their risks of endangerment; mainstreaming biodiversity into the national economic development plan; promoting ecofarming; establishing standardized monitoring techniques; and setting up model areas for well-coordinated biodiversity protection and sustainable development. China has also developed sectoral plans for conservation and the sustainable use of agricultural, forest, marine and coastal, and wetland biodiversity.

China is developing a biodiversity-assessment indicator system and a national-level model of protected areas. To achieve CBD targets, China has also been implementing programs to convert agricultural land back to forests and a licensing system for fishing in marine areas. By 2020, China aims to set up 200 areas for the protection of wild agricultural crops which are the original plant species of China. China has a goal to expand its protected area by up to 18 percent of its total land areas by 2050. The above goals and targets have been included in a number of related plans and programs, such as China's five-year plan for ecology conservation (2006–10) and China's plan for wild animal protection and protected areas. China has established an inter-ministerial mechanism for protecting genetic resources, and has developed a few regulations for the transfer of some genetic resources concerning Protection of New Plant Varieties and the trade of seeds. The country has completed a project on Chinese medicinal herbs for the protection of traditional knowledge (Article 8(j)) related to Chinese herbs. China has also adopted an environmental impact assessment law and Agenda 21, which are designed and implemented to strengthen local communities' capacities in preserving their traditional knowledge and practices. However, China still needs to develop more comprehensive strategies to protect its rich traditional knowledge.

Japan

Japan has 67 percent of its land covered with forest. The length of natural coast is approximately 18,100 km, with 51,500 hectares of tidal flats, 201,200 hectares of Moba (seaweed/seagrass beds), and about 34,700 hectares of coral reefs. These are especially important coastal ecosystems in the context of biodiversity conservation. At least 90,000 species inhabit Japan and its ocean areas.

Japan adopted its third Biodiversity Strategy in 2007. The main goals of this "New Biodiversity Strategy" of Japan are to promote conservation, to prevent species extinction, and restoration of nature throughout the country. These strategy plans are to build up a verdant national land area in which all citizens can enjoy daily interaction with a wide variety of thriving life forms. Some regulations like the Natural Parks Law, the Wildlife Protection and Proper Hunting Law, the Law for the Conservation of Endangered Species of Wild Fauna and Flora, and the Invasive Alien Species Act exist in regard to the objectives of the convention. Regarding marine and coastal areas, Japan has been working to restore about 2,100 hectares of lost marsh and tideland by 2007. As to agriculture, Japan has also taken up environmental conservation such as promoting "Environmentally Conscious Agriculture" and integrating environmental considerations into the implementation of the "Improvement of Agricultural Infrastructure and Rural Development Plan." The country set a target of one hundred percent implementation of mitigation measures to reduce incidental catches of seabirds by 2015. Goals are set to conserve appropriately the biodiversity that is unique to the region. The government of Japan believes that providers should be encouraged to allow smooth access to genetic resources, while users should be encouraged to understand the principles of access to genetic resources and fair and equitable sharing. Based on this principle, in 2005 Japan developed "Guidelines on Access to Genetic Resources for Users in Japan."

European Union

In Europe overall, 971 species are considered globally threatened in the 2003 IUCN red list while 64 endemic plants have already become extinct. At the species level, 42 percent of Europe's native mammals, 43 percent of birds, 45 percent of butterflies, 30 percent of amphibians, 45 percent of reptiles, and 52 percent of freshwater fish are threatened with extinction; many fish stocks are outside safe biological limits (SBL). Some 800 plant species in Europe are at risk of extinction; and there are unknown but potentially significant changes in lower life forms including invertebrate and microbial diversity. Some previously highly threatened species are starting to recover, while others continue to decline at alarming rates, generally as a result of the disappearance or degradation of their habitats. The wetlands in Europe have diminished by up to 60 percent in the last decades. Natural riverside forest used to cover around 2,000 km^2 of areas along the River Rhine. Nowadays, it is highly fragmented and covers only a total of 150 km^2. Threats to biodiversity include land abandonment, urban and transport infrastructure development, climate change, the introduction of invasive alien species, and forest fires.

The European Union Heads of State or Government decided in 2001 "to halt the decline of biodiversity in the EU by 2010" and to "restore habitats and natural systems." The European Commission adopted a new Communication on World Biodiversity Day 2006, which is basically a policy approach aimed at halting the loss of biodiversity by 2010. This approach contains four main policy areas: biodiversity in the European Union (EU), the EU and global biodiversity, biodiversity and climate change, and the knowledge base. It also set out ten primary objectives in relation to biodiversity protection which include the addressing of most important habitats and species; reducing impacts of invasive alien species; actions in the wider countryside and marine environment; making regional development more compatible with nature; support to biodiversity in international development, and so on. Recent reforms of the EU Common Agriculture Policy (CAP) include concerns for biodiversity protection. Key elements of this policy are the introduction of cross-compliance, the promotion of organic farming, support for rural development, and the so-called "agri-environmental measures," a specific European Community program which finances measures to promote the conservation, characterization, collection and utilization of genetic resources, and the adoption of the Biodiversity Action Plan for Biodiversity in Agriculture. The CAP EU Rural Development Policy also aims to reconcile agriculture with the objectives of the CBD, enhancing a series of measures that encourage farmers to protect the landscape and biodiversity, marine, coastal and inland water biodiversity – fisheries. The EU Biodiversity Strategy took up broad objectives for the fisheries sector, and the Biodiversity Action Plan for Fisheries 2001 made recommendations to protect ecosystems of marine fisheries and aquaculture.

The EU gives support to the global initiatives promoting conservation and sustainable use of forest biodiversity, such as the United Nations Forum on Forests. The EU is also a party to the International Tropical Timber Agreement. The EU provides full support to the international legislation concerning endangered species and habitats, including the Convention on International Trade in Endangered Species of Wild Fauna and Flora (CITES). The EU multi-year framework programs for Research and Technological Development (RTD) allocate funds for research in the field of biological diversity conservation, in line with Article 12 of the CBD. The European Platform for Biodiversity Research Strategy (EPBRS) promotes biodiversity research that will contribute to policies and management relating to biodiversity loss. Implementation of the Bonn Guidelines on access to genetic resources and the benefit-sharing guidelines is one of the EU priorities. The EU also provides support for the traditional knowledge of indigenous and local communities.

Conclusion

The CBD is still a controversial issue, and there are arguments among the member countries about their full acceptance of the ecosystem approach of the CBD. The arguments cover three main issues. First, effective implementation of the concept depends on the quality of the approach in terms of its theoretical justification, its consistency, its ability to guide, and its consideration of the existing natural resource management approach currently pursued by most CBD member countries. Second, the implementation of this concept requires the support of the international system beyond the organs of the CBD that serve as adaptors and facilitators for implementation. Third, while the implementation would take place in the national and the sub-national levels, the degree to which the member countries adopt the concept of the ecosystem approach will depend on their institutional, social, and economic capacities as well as political support. However, the world has failed to meet the 2010 target set by the CBD to curb the loss of biodiversity and ecosystems. Various statistics show that initiatives to protect the biodiversity level and ecosystem services are still below what is required to maintain the current level of biodiversity on Earth. Thus the countries should give more stress in different policy adoptions towards the conservation and protection of the biodiversity level and ecosystem services and their proper implementation. Different multilaterals should take up more projects towards these goals too. It has been observed that biodiversity related information is often hidden behind other environmental-related activities, proper information-sharing regarding biodiversity-protection mitigation action among different stakeholders is very important. And as biodiversity protection activities need huge monetary investments, the business sectors should take active parts in these activities and they should also try to include mainstreaming the ecosystem and biodiversity protection initiatives in their core business strategies.

References

Australian Government, 2009, *National Biodiversity Hotspots*, Canberra.

Adams, M. W. and Hutton, J., 2007, "People, parks and poverty: Political ecology and biodiversity conservation," *Conservation Society*, 5(2): 147–183.

African Development Bank, 2010, *Safeguarding Biodiversity and Ecosystem Services: Towards an African Green Economy*, African Development Bank.

African Development Bank, 2011a, *Multi-Resource Forest Inventory for Preparation of a Land Allocation Plan, Republic of Congo*, African Development Bank.

African Development Bank, 2011b, *ADB Projects Supporting Sustainable Forest Management and REDD+*, Manila: Asian Development Bank.

Bhat, G. M., 1996, "Trade-related intellectual property rights to biological resources: Socioeconomic implications for developing countries," *Ecological Economics*, 19(3): 205–217.

Bryant, D., Nielsen, S. and Tangley, L., 1997, *The Last Frontier Forests: Ecosystems and Economies on the Edge*, Seattle, WA: World Resource Institute.

Convention on Biological Diversity, 2011a, *History of the Convention*, Secretariat of the Convention on Biological Diversity, 2004, Biodiversity Issues for Consideration in the Planning, Establishment and Management of Protected area sites and networks. Montreal: Secretariat of the Convention on Biological Diversity, 164 pages and i to iv, CBD Technical Series no. 15.

Convention on Biological Diversity, 2011b, *The Cartegena Protocol on Biosafety*, Secretariat of the Convention on Biological Diversity, 2000, Cartagena Protocol on Biosafety to the Convention on Biological Diversity: Text and annexes, Montreal: Secretariat of the Convention on Biological Diversity.

Crucible Group, 1994, *People, Plants and Patents: The Impact of Intellectual Property on Trade, Plant Biodiversity, and Rural Society*, Ottawa: International Development Research Centre.

Department of Sustainability, Environment, Water, Population and Communities (DSEWPC), 2009, Megadiverse Countries, Government of Australia [cited 2011].

Escobar, A., 1998, "Whose knowledge, whose nature? Biodiversity, conservation, and the political ecology of social movements," *Journal of Political Ecology*, 5(1): 53–82.

European Environment Agency, 2011, *EEA Homepage on Biodiversity*, Denmark: EEA.

Fisher, B. and Christopher, T., 2007, "Poverty and biodiversity: Measuring the overlap of human poverty and the biodiversity hotspots," *Ecological Economics*, 62(1): 93–101.

Global Environmental Facility, 2004, *GEF and the Convention on Biological Diversity: A Strong Partnership with Solid Results*, Washington DC: Global Environmental Facility.

G/Egziabher, T. B., 2000, *Intellectual Property Rights in Biological Diversity and Trade Agreements*, Addis Ababa: The Institute for Sustainable Development.

Hartje, V., Klaphake, A. and Schliep, R., 2003, *The International Debate on the Ecosystem Approach: Diffusion of a Codification Effort,* Working Paper, Institute for Landscape and Environmental Planning.

Hoekstra, J. M., Boucher, T. M., Ricketts, T. H. and Roberts, C., 2005, "Confronting a biome crisis: Global disparities of habitat loss and protection," *Ecology Letters* 8(1): 23–29.

Lele, S., Wilshusen, P., Brockington, D., Seidler, R. and Bawa, K., 2010, "Beyond exclusion: Alternative approach to biodiversity conservation in the developing tropics," *Current Opinion in Environmental Sustainability*, 2(1): 1–7.

Medaglia, J. C., 2010, *The Political Economy of the International ABS Regime Negotiations*, Geneva: International Centre for Trade and Sustainable Development.

Millennium Ecosystem Assessment, 2005, *Ecosystems and Human Well-being: Synthesis*, Washington, DC: Island Press.

Mittermeier, R. A., Robles-Gil, P., Mittermeier, C. G. (eds), 1997, *Megadiversity: Earth's Biologically Wealthiest Nations*, CEMEX/Agrupaciaon, Sierra Madre, Mexico City.

Mittermeier, R. A., Mittermeier, C. G., Brooks, T. M., Pilgrim, J. D., Konstant, W. R., da Fonseca G. A. B. and Kormos, C., 2003, "Wilderness and biodiversity conservation," *Proceedings of the National Academy of Sciences*, 100: 10309–10313.

Neumann, P. R., 2004, "Moral and discursive geographies in the war for biodiversity in Africa," *Political Geography*, 23(7): 813–837.

Olson, D. M. and Dinerstein, E., 1998, "The Global 200: A representation approach to conserving the Earth's most biologically valuable ecoregions," *Conservation Biology*, 12(3): 502–515.

Sanderson, E. W., Redford, K. H., Vedder, A., Coppolillo, P. B. and Ward, S. E., 2002, "A conceptual model for conservation planning based on landscape species requirements," *Landscape and Urban Planning*, 58(1): 41–56.

Stattersfield, A. J., Crosby, M. J., Long, A. J. and Wege, D. C., 1998, "Endemic bird areas of the world: Priorities for biodiversity conservation," *Birdlife Conservation Series*, 7, Cambridge Birdlife International.

The Gallup Organization, 2010, *Attitudes of Europeans Towards the Issue of Biodiversity*, Directorate General Environment, European Commission.

Turner, W. R., Brandon, K., Brooks, M. T., Costanza, R., Fonseca, G. and Portela, R., 2007, "Global conservation of biodiversity and ecosystem services," *BioScience*, 57(10): 868.

Velásqueznts, G. and Boulet, P., 1999, *Globalization and Access to Drugs: Implications of the WTO/TRIPS Agreement*, Geneva: World Health Organization.

Warren, D. M., Slikkerveer, L. J. and Brokensha, D. (eds), *The Cultural Dimension of Development: Indigenous Knowledge Systems*, London: Intermediate Technology Publications: 479–487.

Wilson, Edward O., 1992, "The Diversity of Life," reviewed by Garrett Hardin, *Population and Development Review*, 19(1): 183–186.

World Bank, 2008, *Biodiversity, Climate Change and Adaptation: Nature-based Solutions from the World Bank Portfolio*, Washington, DC: The World Bank.

World Bank, 2010, "The role of biodiversity and ecosystems in sustainable development," *2010 Environment Strategy*, Washington, DC: The World Bank Group.

Part II
Valuation of ecosystem services

3 The value of biodiversity and recreation demand models

A spatial Kuhn–Tucker model

Koichi Kuriyama, Yasushi Shoji, and Takahiro Tsuge

Introduction

Ecosystem services are public goods with no market price. In this chapter, we apply the recreation demand model to estimate the value of biodiversity. Biodiversity provides an array of recreational opportunities. For example, the Yakushima World Heritage site, Southern Japan has more than 1,900 species and the unique ecosystem and natural beauty of the site attract more than 200,000 visitors per year. If the biodiversity of Yakushima was lost, most of the visitors might cancel their tour to the site. Thus, the damage to the biodiversity might be estimated using the monetary loss of recreational expenditure.

Recreation sites have characteristics of spatial data. For example, the geographic attributes of recreation sites, such as altitude, temperature, vegetation, and land use are distributed spatially. Thus, recreation demand has spatial heterogeneity due to the spatial characteristics of the recreation sites.

For demand analysis of spatial data, spatial econometric approaches have been developed. Spatial econometrics is a subfield of econometrics that deals with the treatment of spatial interaction (spatial autocorrelation) and spatial structure (spatial heterogeneity) in regression models (Anselin 1988). Spatial autoregression models are commonly used in spatial econometrics to analyze spatial autocorrelation (Anselin 1988). For spatial heterogeneity, a spatial error component model has been proposed (Kelejian and Robinson 1993, 1995, 1997). In the environmental economics literature, a spatial hedonic approach using explicit spatial regression techniques has been developed. Examples of empirical studies using the spatial hedonic approach to estimate the effect of environmental quality include Leggett and Bockstael (2000) on water quality, Acharya and Bennett (2001) on open spaces, Paterson and Boyle (2002) on visibility, Kim *et al.* (2003) on air quality, Cho *et al.* (2006) on water and green spaces, Kim and Goldsmith (2009) on swine production, and Hoshino and Kuriyama (2010) on urban parks amenities. However, no empirical study has used the spatial econometric approach to analyze recreation demand.

Recreation demand is characterized by a mixture of corner and interior solutions. For a given individual, many sites are unvisited (corner solutions), while other sites are visited once or more (interior solutions). To deal with this, previous studies have used the Kuhn–Tucker (KT) model. This model relies on a single structural framework to model site selection and participation decisions simultaneously, allowing for corner solutions. The KT model was initially proposed by Hanemann (1978) and Wales and Woodland (1983) and later refined by Bockstael *et al.* (1986). Phaneuf *et al.* (2000) (PKH) applied this model to recreation demand. The PKH strategy for the analytical calculation of compensating variation is not feasible when the choice set is relatively large; however, von Haefen *et al.* (2004) proposed a bisection algorithm for constructing compensating variation with a large choice dataset.

Recent studies of recreation demand have found heterogeneity in preferences (Phaneuf and Smith 2006; Hilger and Hanemann 2006). For example, beach recreation activities include swimming, fishing, surfing, and biking. People engaged in beach recreation who swim are likely to have different preferences for water quality than do those who relax on the sand. Thus, demand for beach recreation may be heterogeneous. To capture unobserved heterogeneity in a KT framework, recent work proposed a random parameter model, using maximum simulated likelihood (von Haefen *et al.* 2004) and Bayesian estimation (von Haefen and Phaneuf 2005). Recently, Kuriyama *et al.* (2010) proposed the latent segmentation approach to the KT model; this accounts for heterogeneity in preferences and provides information regarding preference composition.

In this chapter, we extend the spatial econometric approach to the KT model. The proposed approach models spatial heterogeneity using a spatial error component that varies across sites. Our proposed approach is appealing in that it can analyze both spatial and non-spatial heterogeneity within a single structural KT framework, which simultaneously models recreational participation and site-selection decisions, while allowing for corner solutions and consistency with utility theory.

The remainder of the chapter is organized as follows. The next section outlines our spatial econometric approach to the KT model. The third section gives an empirical illustration of the proposed model using data for park recreation in Hokkaido, Japan. The final section contains concluding comments.

Model

Let the consumer's direct utility function be $U(\mathbf{x}, \mathbf{Q}, z, \boldsymbol{\beta}, \mathbf{E})$, where $\mathbf{x} = (x_1, \ldots, x_M)'$ is a vector of visits to the recreation sites, $\mathbf{Q} = (\mathbf{q}_1, \ldots, \mathbf{q}_M)'$ is an $M \times K$ matrix of quality attributes for the M sites, z is the Hicksian composite good, $\boldsymbol{\beta}$ is an vector of parameters, and $\mathbf{E}$ is a vector or matrix of unobserved heterogeneity. The spatial error-component model assumes that $\mathbf{E}$ contains two uncorrelated error components $\boldsymbol{\eta}$ and $\boldsymbol{\varepsilon}$, such that

$$\mathbf{E} = \mathbf{W}\boldsymbol{\eta} + \boldsymbol{\varepsilon}$$

where η is the unobserved heterogeneity across sites and ε is the unobserved heterogeneity across sites and individuals. $\mathbf{W}$ is an $M \times M$ spatial weight matrix, which has the following elements:

$$w_{jk} = \begin{cases} \dfrac{1}{d_{jk}}, & j \neq k \\ 0, & j = k \end{cases}$$

where d_{jk} is the distance between sites j and k. For ease of interpretation, the elements of the spatial weight matrix are row-normalized such that all rows sum to 1. Thus, the term $\mathbf{W}\eta$ indicates a smoothing of neighboring values or a regional effect, and the term ε is site-specific.

The individual is assumed to maximize the utility function subject to budget and non-negativity constraints:

$$\text{Max} \quad U(\mathbf{x}, \mathbf{Q}, z, \boldsymbol{\beta}, \mathbf{E})$$
$$\text{s.t.} \quad \mathbf{p}'\mathbf{x} + z \leq y, \quad z > 0, \mathbf{x} \geq 0 \tag{3.1}$$

where U is assumed to be a quasi-concave, increasing, and continuously differentiable function of $(\mathbf{x}, \mathbf{z})$, $p = (p_z, \ldots, p_M)'$ is a vector of prices, and y is the income. The first-order KT condition for the problem is as follows:

$$\frac{\partial U}{\partial x_j} \leq p_j \cdot \frac{\partial U}{\partial z}$$
$$x_j^* \geq 0,$$
$$x_j^* \left[\frac{\partial U}{\partial x_j} - p_j \frac{\partial U}{\partial z} \right] = 0 \tag{3.2}$$

for $j = 1, \ldots, M$, where $\mathbf{x}^*$ is a solution vector of the problem (3.1). Following PKH, we assume that $\partial^2 U/\partial z \partial \varepsilon = 0$, $\partial^2 U/\partial x_j \partial \varepsilon_k = 0$ for $\forall k \neq j$, and $\partial^2 U/\partial x_j \partial \varepsilon_k > 0$ for $\forall j$. Under this assumption, the first-order condition can be rewritten as

$$\varepsilon_j \leq g_j - \mathbf{W}\eta_j$$
$$x_j^* \geq 0$$
$$x_j^* [g_j - \mathbf{W}\eta_j - \varepsilon_j] = 0$$

for $j = 1, \ldots, M$, where g_j is the solution to

$$\frac{\partial U(\mathbf{x}^*, \mathbf{Q}, y - \mathbf{p}'\mathbf{x}^*, \boldsymbol{\beta}, g_j)}{\partial x_j} = p_j \cdot \frac{\partial U(\mathbf{x}^*, \mathbf{Q}, y - \mathbf{p}'\mathbf{x}^*, \boldsymbol{\beta}, g_j)}{\partial z}.$$

Assume that ε_j is an independent and identically distributed draw from the type-I extreme value distribution with inverse scale parameter μ for all j. The probability of observing $\mathbf{x}$, conditional on the spatial error term, is as follows:

$$l(\mathbf{x} \mid \boldsymbol{\eta}) = |\mathbf{J}| \prod_j G(-g_j + \mathbf{W}\boldsymbol{\eta}_j)$$

where

$$G(-g_j + \mathbf{W}\boldsymbol{\eta}_j) = \left[\frac{1}{\mu} \exp\left(\frac{-g_j + \mathbf{W}\boldsymbol{\eta}_j}{\mu} \right) \right]^{1[x_j^* > 0]} \exp\left[-\exp\left(\frac{-g_j + \mathbf{W}\boldsymbol{\eta}_j}{\mu} \right) \right].$$

$|\mathbf{J}|$ is the determinant of the Jacobian for the transformation, and $1[x_j^* > 1]$ is an indicator function equal to 1 if x_j^* is strictly positive and zero otherwise. The unconditional probability can be calculated by

$$l(\mathbf{x}) = \int l(\mathbf{x} \mid \boldsymbol{\eta}) f(\boldsymbol{\eta}) \, d\boldsymbol{\eta}$$

$$= |\mathbf{J}| \int \prod_j G(-g_j + \mathbf{W}\boldsymbol{\eta}_j) f(\boldsymbol{\eta}) \, d\boldsymbol{\eta}_j, \tag{3.3}$$

where $f(\boldsymbol{\eta})$ is a density function. Note that $|\mathbf{J}|$ can be pulled outside the integral in Equation (3.3) because the elements of the Jacobian ($\partial \varepsilon_j / \partial x_k$) do not include the spatial error $\boldsymbol{\eta}$. The probability can be approximated through simulation: (1) draw a value of $\boldsymbol{\eta}$ from $f(\boldsymbol{\eta})$, and label it $\boldsymbol{\eta}^r$ with the superscript $r = 1, \ldots, R$ referring to the rth draw; (2) calculate the probability $l(\mathbf{x} \mid \boldsymbol{\eta}^r)$ with this draw; then (3) repeat steps 1 and 2 and average the results. This average is the simulated probability:

$$\tilde{l}(\mathbf{x}) = \frac{1}{R} \sum_{r=1}^{R} l(\mathbf{x} \mid \boldsymbol{\eta}^r)$$

$$= \frac{|\mathbf{J}|}{R} \sum_{r=1}^{R} \prod_j G(-g_j + \mathbf{W}\boldsymbol{\eta}_j).$$

Note that it is not necessary to calculate $|\mathbf{J}|$ for each simulation step when simulating $l(\mathbf{x})$ because $|\mathbf{J}|$ is pulled outside the summation.

Preference specification and welfare analysis

Following PKH, we used the additively separable utility function:

$$U(\cdot) = \sum_{j}^{M} \Psi_j \ln(x_j + \theta_j) + \frac{1}{\rho} z^{\rho}$$

$$\ln \Psi_j = \boldsymbol{\gamma}' \mathbf{q}_j + \mu \varepsilon_j + \sigma \mathbf{W} \boldsymbol{\eta}_j$$

$$\rho = 1 - \exp(\rho^*)$$

$$\ln \theta = \theta^*$$

$$\ln \mu = \mu^*, \tag{3.4}$$

where $(\boldsymbol{\gamma}, \sigma, \theta^*, \rho^*, \mu^*)$ are parameters. We assume that η is an independent and identically distributed draw from the normal distribution, with standard deviation parameter σ for all j. As theoretical constraints require $\rho \leq 1$, $\theta > 0$, and $\mu > 0$, we used a transformation of parameters, that is $\rho = 1 - \exp(\rho^*)$, $\ln \theta = \theta^*$, and $\ln \mu = \mu^*$. Because the utility function in Equation (3.4) is not consistent with the weak complementarity, except for the special case of $\theta = 1$, this specification may be suitable for the case in which the values of recreation sites include non-use values.

The first-order condition of the utility maximization problem can be written as

$$\varepsilon_j \leq \frac{1}{\mu}[-\boldsymbol{\gamma}' \mathbf{q}_j - \sigma \mathbf{W} \boldsymbol{\eta}_j + \ln p_j + \ln(x_j + \theta_j) + (\rho - 1) \ln z], \tag{3.5}$$

for all j. Assume that the parameters of the site's quality attributes vary randomly across individuals. From Equation (3.5), $\partial \varepsilon_j / \partial x_k$ does not include a random parameter $(\boldsymbol{\gamma})$; therefore, it is not necessary to calculate $|\mathbf{J}|$ for each simulation step when simulating the probability, because $|\mathbf{J}|$ is not affected by the random parameters $(\boldsymbol{\gamma})$. Given this significant computational saving, it is tractable to calculate the simulated probability for capturing the unobserved heterogeneity using the random parameters varying across individuals, as well as the spatial heterogeneity in the spatial error components that vary across sites.

Let $v(\cdot)$ be the indirect utility function and $e(\cdot)$ be the expenditure function. The Hicksian compensating variation CV for a change in prices or quantities from $(\mathbf{p}^0, \mathbf{q}^0)$ to $(\mathbf{p}^1, \mathbf{q}^1)$ can be defined using the indirect utility function

$$v(\mathbf{p}^0, \mathbf{q}^0, y, \boldsymbol{\beta}, \mathbf{E}) = v(\mathbf{p}^1, \mathbf{q}^1, y - CV, \boldsymbol{\beta}, \mathbf{E})$$

or using the expenditure function

$$CV = y - e(\mathbf{p}^1, \mathbf{q}^1, U^0, \boldsymbol{\beta}, \mathbf{E})$$

where $U^0 = v(\mathbf{p}^0, \mathbf{q}^0, y, \boldsymbol{\beta}, \mathbf{E})$. Unless the preferences are homothetic or quasi-linear in income, no closed-form solution for the CV exists; consequently, this situation requires an iterative search procedure. However, when the dimension of the choice set is large, solving CV analytically, as done by PKH, is not feasible. von Haefen *et al.* (2004) proposed an iterative algorithm that numerically solves the consumer's constrained maximization problem at each iteration of the numerical bisection method. von Haefen (2003) developed a more efficient numerical algorithm that relies on expenditure functions, using the consumer's constrained minimization problem. Using the procedure of von Haefen (2003), and assuming additively separable utility, the following algorithm finds the CV:

(1) At iteration i, set $z_a^i = (z_l^i + z_u^i)/2$. To initialize the algorithm, set $z_l^0 = 0$ and z_u^0 to a solution to $U(0, \mathbf{Q}^1, z_u^0, \boldsymbol{\beta}, \mathbf{E}) = U^0$.
(2) Conditional on z_a^i, solve for $\mathbf{x}^i$ using the first-order condition (3.2). Calculate $U^i = U(\mathbf{x}^i, \mathbf{Q}^1, z_a^i, \boldsymbol{\beta}, \mathbf{E}) = U^0$.
(3) If $U^i < U^0$, set $z_l^i = z_a^i$ and $z_u^i = z_l^i$. Otherwise set $z_l^i = z_l^i$ and $z_u^i = z_a^i$.
(4) Iterate until $|z_l^i - z_u^i| < c$, where c is arbitrarily small.
(5) Calculate $CV = y - (\mathbf{px}^i + z_a^i)$.

Note that the compensating variation CV has random components; thus, it is impossible to estimate the exact value of CV. However, as shown by von Haefen *et al.* (2004), the expectation of the compensating variation CV can be estimated using first-order conditions and a simulation technique for the unobserved components. To simulate unobserved heterogeneity for the preference parameter $\boldsymbol{\beta}$, and for the error components $(\boldsymbol{\varepsilon}, \boldsymbol{\eta})$, we need to draw from the joint distribution $f(\boldsymbol{\beta}, \boldsymbol{\varepsilon}, \boldsymbol{\eta} \mid \mathbf{x})$ conditional on the observed trip demand. The joint distribution can be decomposed as:

$$f(\boldsymbol{\beta}, \boldsymbol{\varepsilon}, \boldsymbol{\eta} \mid \mathbf{x}) = f(\boldsymbol{\beta}, \boldsymbol{\eta} \mid \mathbf{x}) f(\boldsymbol{\varepsilon} \mid \boldsymbol{\beta}, \boldsymbol{\eta}, \mathbf{x}).$$

Following von Haefen *et al.* (2004), we use an adaptive Metropolis-Hastings algorithm to simulate from $f(\boldsymbol{\beta}, \boldsymbol{\eta} \mid \mathbf{x})$.[1] Note that the random parameters in $\boldsymbol{\beta}$ vary across the population and require N draws for all N observations, while the spatial error $\boldsymbol{\eta}$ varies across sites and requires M draws for all M sites. After a burn-in period, this algorithm generates random draws from $f(\boldsymbol{\beta}, \boldsymbol{\eta} \mid \mathbf{x})$. For the simulation of random draws from $f(\boldsymbol{\varepsilon} \mid \boldsymbol{\beta}, \boldsymbol{\eta}, \mathbf{x})$, if site j is visited, the first-order condition implies that $\varepsilon_j = (g_j - \sigma \mathbf{W} \boldsymbol{\eta}_j)/\mu$. Otherwise, estimates of $\varepsilon_j < (g_j - \sigma \mathbf{W} \boldsymbol{\eta}_j)/\mu$ can be simulated, given the type-I extreme-value distribution assumptions. For details of the simulation of the unobserved components for the KT model, see von Haefen and Phaneuf (2005).

Empirical illustration

To illustrate our proposed model with spatial heterogeneity, we used data on recreation trips to the national parks and quasi-national parks[2] in Hokkaido, the northern island of Japan (Figure 3.1). There are six national parks and five

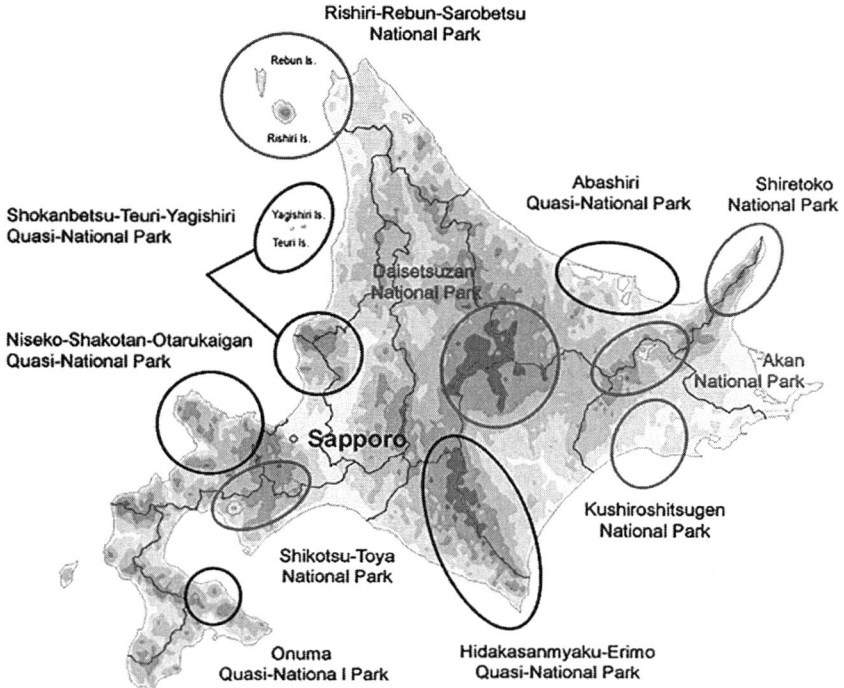

Figure 3.1 National and quasi-national parks in Hokkaido, Japan.

quasi-national parks in Hokkaido, including Shiretoko, a World Heritage site. Table 3.1 summarizes the characteristics of these parks. All of the parks include protected species of vegetation. For some, there is restricted road access to the parks by private car users because of the negative environmental impact of traffic jams.

Figure 3.2 shows the altitude of Hokkaido. Most of the parks are located in mountain areas, except Kushiro Shitsugen and Abashiri. Figure 3.3 shows the average temperature. This suggests that the temperature of the parks has spatial characteristics and it may be a source of the spatial heterogeneity. Figure 3.4 shows the number of species. Additionally, it suggests that the spatial heterogeneity is caused by the spatial distribution of biodiversity.

To investigate the number of trips to each national and quasi-national park in Hokkaido from the end of July to the end of November in 2008, a web survey was conducted in 2008 between 5 December and 9 December. Invitations to complete the survey were sent to 3,052 general residents in Hokkaido by email and 763 people responded.

Table 3.1 Characteristics of National Parks in Hokkaido, Japan

Site ID	Name of park	Park type[1]	Total area (ha)	Protected species of vegetation (No. of species)	Car restriction (km)
1	Rishiri-Rebun-Sarobetsu	NP	24,166	263	0
2	Daisetsuzan	NP	226,764	291	25
3	Shikotsu-Toya	NP	99,473	215	1.8
4	Shiretoko	NP	38,633	245	11
5	Akan	NP	90,481	198	0
6	Kushiro Shitsugen	NP	26,861	158	0
7	Shokanbetsu-Teuri-Yagishiri	QNP	43,559	202	0
8	Abashiri	QNP	37,261	88	0
9	Niseko-Shakotan-Otarukaigan	QNP	19,009	171	0
10	Hidaka-sanmyaku-Erimo	QNP	103,447	396	0
11	Onuma	QNP	9,083	132	0

Note: NP: national park, QNP: quasi-national park.

Source: The Statistical Handbook of Environment, Ministry of the Environment (2011).

Table 3.2 describes the site quality attributes and other variables used in our empirical analysis. The travel cost was calculated as

$$TC = \frac{Dist(km) \times 2}{10(km/1)} \times 100(yen/1) + Turnpike(yen) \times 2$$
$$+ Time(hour) \times 2 \times 2803(yen/hour) \times \frac{1}{3},$$

where *Dist* is the one-way travel distance between the respondent's residence and each site in the choice set. The price of gas was assumed to be 100 yen per liter and the fuel mileage was assumed to be 10 km per liter. *Turnpike* is one-way turnpike tolls. *Time* is the one-way travel time. The opportunity cost per hour is assumed to be one-third of the average wage rate (2,803 yen) in 2000 (Cesario 1976).

Multi-destination trips were included because they might be a cause of the spatial behavior in recreation demand. We recoded multi-destination trips as single-destination trips. For example, the multi-destination trip path "home-site A-site B-home" was recoded as two trips: one to site A and one to site B.

Figure 3.5 shows the relationship between the average one-way trip cost and the average number of trips taken per respondent. This suggests that the demand for sites 3 (Shikotsu-Toya) and 11 (Onuma) was relatively high, while that for sites 7 (Shokanbetsu-Teuri-Yagishiri) and 10 (Hidaka-sanmyaku-Erimo) was relatively low, compared with the others.

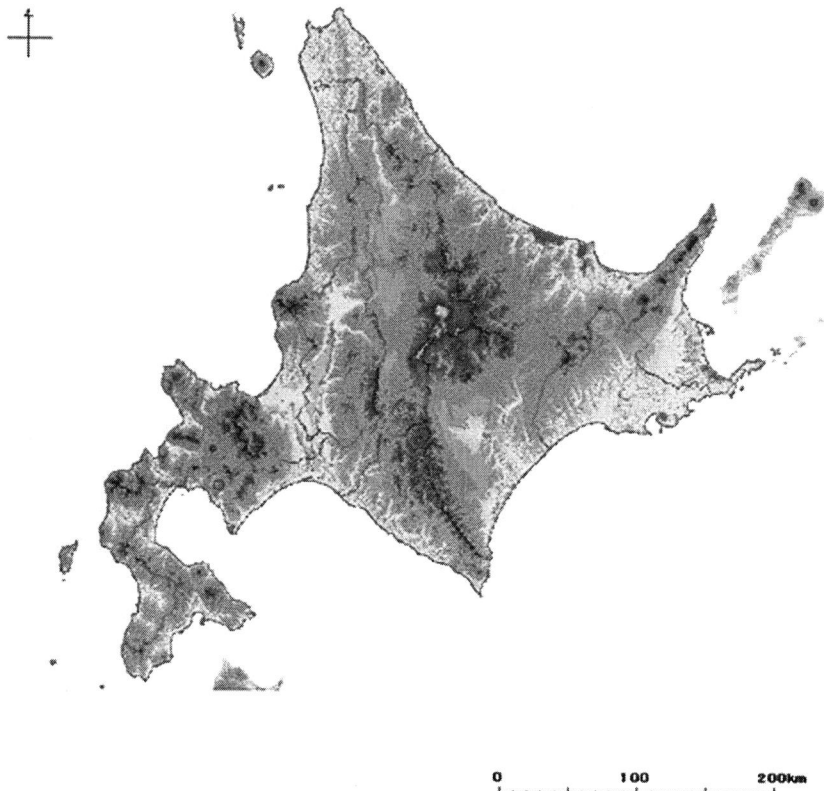

Figure 3.2 Altitude in Hokkaido, Japan.

Source: GIS data from the Resources and Environment Database, Hokkaido Research Organization, Institute of Environmental Science, Hokkaido, Japan, http://envgis.ies.hro.or.jp/

Table 3.2 Variable list

Name	Mean	Std. dev.	Note
TC	18,748	9,926	Two-way travel cost
Trips	0.874	1.580	Number of trips taken per respondent
Area	6.534	6.306	Total area of parks (10,000 ha)
Species	2.145	0.839	Number of protected species of vegetation (100 species)
Car Restriction	0.034	0.079	The distance of car restriction road (100 km)

Source: Data survey conducted by the authors.

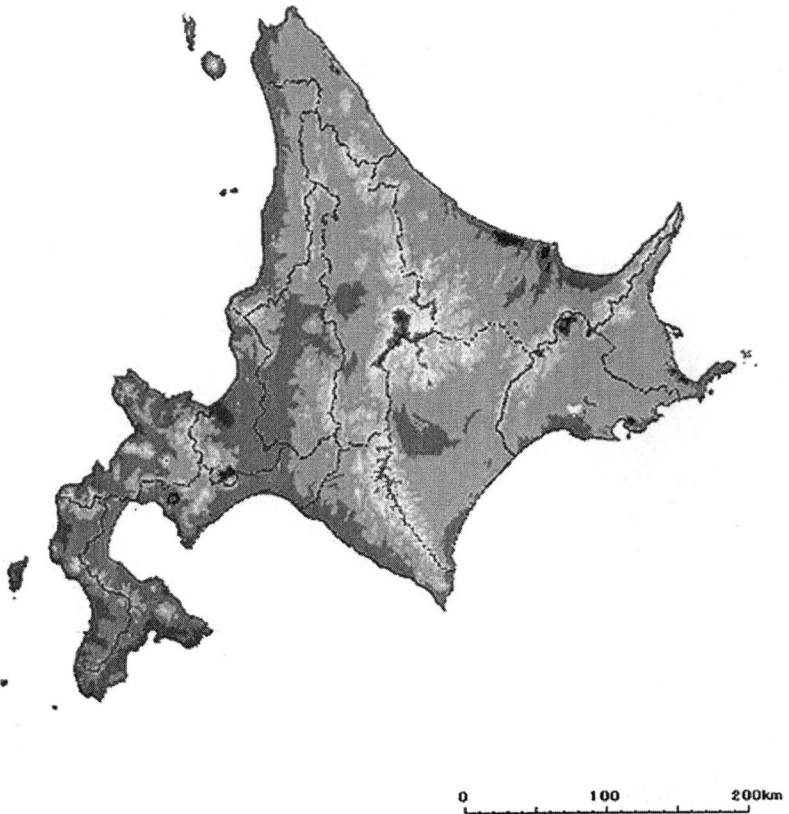

Figure 3.3 Average temperature in Hokkaido, Japan.

Source: GIS data from the Resources and Environment Database, Hokkaido Research Organization, Institute of Environmental Science, Hokkaido, Japan, http://envgis.ies.hro.or.jp/

Estimation results

We compared three models: (1) the standard KT model; (2) the spatial KT model with fixed parameters; and (3) the spatial KT model with random parameters. For the standard KT model, the parameters were assumed to be equal across the population; thus, it is difficult to capture both unobserved heterogeneity and spatial heterogeneity, except for the independent random component. For this model, all of the parameters were significant except for the constant term. We use dummy

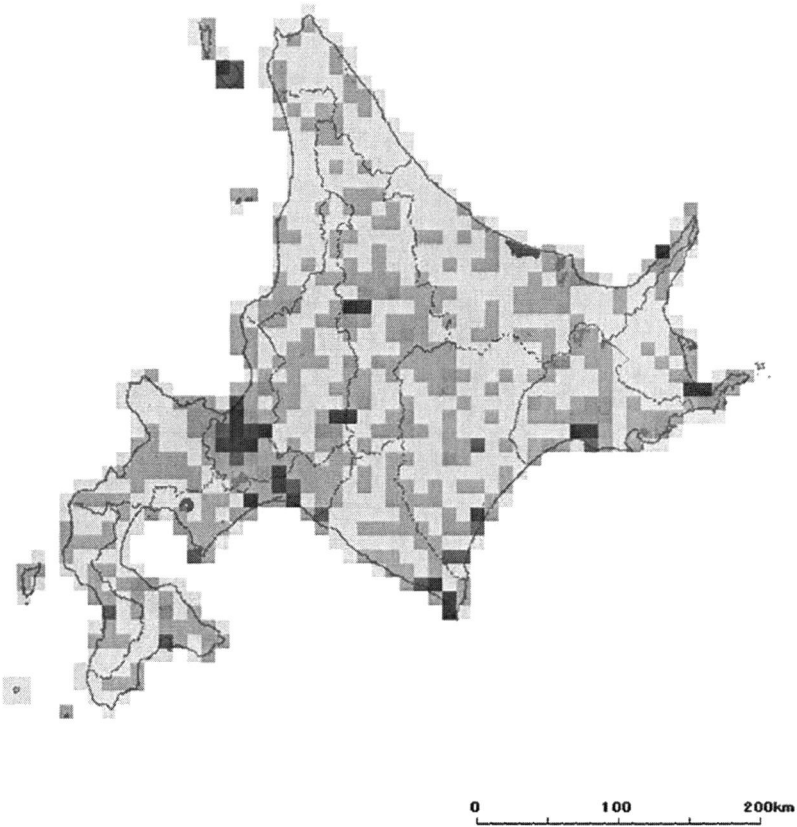

Figure 3.4 The number of species in Hokkaido, Japan.

Source: GIS data from the Resources and Environment Database, Hokkaido Research Organization, Institute of Environmental Science, Hokkaido, Japan, http://envgis.ies.hro.or.jp/

variables for sites 7, 10, and 11, while site 3 was removed because we found that it was not significant. As expected from Figure 3.5, the dummy variables site 7 and site 10 had significantly negative effects on utility, while site 11 had a positive sign. Of the site quality attributes, Area and Species take on positive signs, while Car Restriction had a negative impact on utility.

The spatial KT model with fixed parameters explicitly accounts for spatial heterogeneity, although the parameters are assumed to be fixed across individuals. The parameter spatial weight (σ) is positive and significantly different from zero,

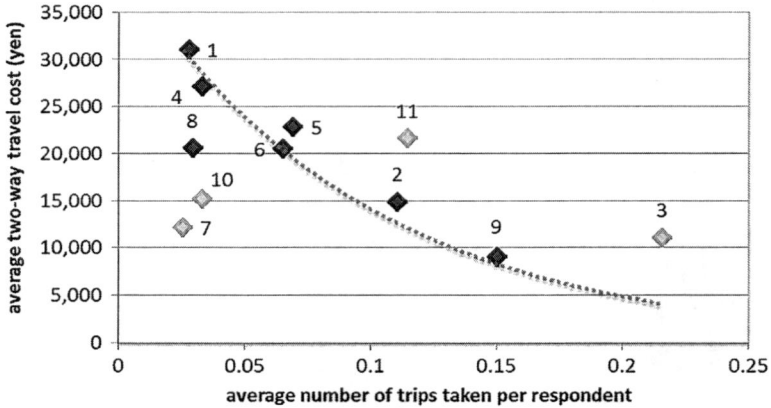

Figure 3.5 Travel cost and number of trips.

Source: Data from a survey conducted by the authors.

so there is spatial heterogeneity in our dataset on recreation demand. For the spatial KT model with random parameters, the random parameters were assumed to vary across individuals. The parameters of the random-parameter KT model were estimated by maximum simulated likelihood with 100 Halton draws.[3] The parameters for the standard deviation of the random parameters of Site 7, Site 11, and Area were significant. Additionally, spatial weight was positive and significant. Thus, our dataset has both spatial heterogeneity across sites and non-spatial heterogeneity across individuals.

Welfare analysis

We considered the following three scenarios to deal with policy implications for recreation in the parks in Hokkaido: (a) close Shiretoko National Park, a World Heritage site; (b) increase the car restriction by 50 percent; and (c) decrease the protection of species by 50 percent. Regarding scenario (a), Shiretoko National Park is closed in some seasons because the site is a natural habitat of the brown bear and the park manager needs to prevent accidents due to encounters between visitors and bears. For scenario (b), road congestion in the parks had a negative impact on the natural environment of the parks. While the Ministry of the Environment has regulated traffic in the parks, the decrease in visitors might have negative effects on the local economy around the parks. For scenario (c), there are more than 100 of the protected plant species at the national and quasi-national parks in Hokkaido. If the endangered species were lost, some tourists might decide against visiting a site.

Table 3.4 shows the compensating variation for the three scenarios. Even if the spatial weight is significant, differences in the compensating variation between

Table 3.3 Estimated parameters

	Model 1 fixed parameters	Model 2[d] fixed parameters	Model 3[d] mean	Std. dev.
Parameters				
const.	0.8793	1.0760	1.3182	
	(1.14)	(1.27)	(1.27)	
Site 7	−1.1969***	−1.1650***	−2.3007**	1.3929**
	(0.23)	(0.21)	(1.06)	(0.63)
Site 10	−1.5993***	−1.4863***	−1.4760***	0.2470
	(0.33)	(0.30)	(0.48)	(1.00)
Site 11	0.3581***	0.4993***	−0.0660	1.1974***
	(0.13)	(0.13)	(0.24)	(0.20)
Area	0.0487***	0.0422***	0.0342***	0.0218**
	(0.01)	(0.01)	(0.01)	(0.01)
Species	0.2884**	0.2921**	0.3033***	0.0454
	(0.13)	(0.12)	(0.12)	(0.06)
Car Restriction	−3.6798***	−3.1110***	−3.2731***	1.0581
	(0.97)	(0.89)	(0.91)	(1.08)
Spatial weight (σ)		0.8677***	0.7763***	
		(0.07)	(0.08)	
Additional parameters				
ρ	−0.9634***	−1.0239***	−1.0278***	
	(0.18)	(0.22)	(0.22)	
θ	1.3390***	1.3515***	1.3794***	
	(0.07)	(0.07)	(0.07)	
μ	−0.2782***	−0.3787***	−0.465***	
	(0.04)	(0.04)	(0.04)	
Log-likelihood	−2393.81	−2346.32	−2338.86	
AIC	4807.62	4714.63	4711.71	
AIC3	4817.62	4725.63	4728.71	
BIC	4853.85	4765.48	4790.30	

[a] $N = 752$.
[b] The values in parentheses are standard errors.
[c] *, ** and *** denote significance at the 10, 5 and 1% levels, respectively.
[d] Parameter estimates generated with 100 Halton draws.

the standard KT model and the spatial KT models were relatively small. Also, this table shows that the decreasing the protected species causes greater damage than the closure of the Shiretoko World Heritage site. Thus, these results suggest that it is important to protect the biodiversity of the national parks.

Concluding comments

We considered a spatial econometric approach to the KT model. Our proposed approach is appealing in that we can analyze spatial heterogeneity within the

Table 3.4 Compensating variations

	Model 1 standard KT	Model 2[c] spatial KT fixed parameters	Model 3[c] spatial KT with random parameters
(a) Closing Shiretoko	−285.85 (15.06)	−282.90 (14.71)	−276.63 (15.91)
(b) Increasing car restriction by 50%	−167.82 (30.18)	−147.45 (29.67)	−143.59 (35.55)
(c) Decreasing protection of species by 50%	−2846.90 (882.22)	−2851.54 (724.49)	−2854.41 (707.65)

[a] Unit: yen per respondent; US$1 = 103 yen in 2008.
[b] The values in parentheses are standard errors calculated using the procedure reported by Krinsky and Robb (1986) and based on 300 iterations.
[c] Welfare estimates generated with 1,250 draws; the first 250 were discarded as burn-in, and every tenth iteration thereafter was used to construct the reported estimates.

single-structure KT framework, which simultaneously models recreational participation and site-selection decisions, while allowing for corner solutions and consistency with utility theory. The proposed and standard KT models were applied to a recreation dataset for the national and quasi-national parks in Hokkaido, Japan, and the model results were compared. The empirical analysis showed that both spatial heterogeneity across sites and non-spatial heterogeneity across individuals exist. For welfare analysis, we considered three scenarios. Estimated results show that decreasing the protected species has a larger welfare loss than does the one of site closure of the Shiretoko World Heritage site. The results suggest that it is important to account for the value of biodiversity in park management.

We suggest two additional paths of investigation to the spatial approach to the KT model. First, other spatial econometric approaches to the KT model should be investigated. In the spatial econometrics literature, some models have been developed for spatial data, such as the spatial lag model and the spatial autoregressive model. The application of these models to the KT framework may be meaningful. However, it might be complex because the utility function used in the KT model is not linear with respect to the error term. Second, multi-destination trips should be analyzed. Visitors taking multi-destination trips may prefer neighboring sites to remote sites. Thus, multi-destination trips may cause spatial behavior in recreation demand. We used a simple approach to the multi-destination trips. Nevertheless, a sophisticated analysis of the multi-destination trips, particularly for the spatial KT model, is an important issue for future research.

Notes

1 Train (2009) illustrates the random draw using the Metropolis-Hastings algorithm.
2 While the national parks are managed by the Ministry of the Environment, the quasi-national parks are managed by the prefectural government.

3 Bhat (2001) found that 100 Halton draws provided more precise results than 1,000 random draws.

References

Acharya, G. and Bennett, L. L., 2001, "Valuing open space and land-use patterns in urban watersheds," *The Journal of Real Estate Finance and Economics*, 22(2–3): 221–237.

Anselin, L., 1988, *Spatial Econometrics: Methods and Models*, Dordrecht: Kluwer Academic.

Bhat, C., 2001, "Quasi-random maximum simulated likelihood estimation of the mixed multinomial logit model," *Transportation Research B*, 35(7): 677–693.

Bockstael, N., Hanemann, W. M. and Strand, I. E., 1986, *Measuring the Benefits of Water Quality Improvements Using Recreation Demand Models*, Vol. 2, Washington DC: US Environmental Protection Agency, Office of Policy Analysis.

Cesario, F. J., 1976, "Value of time in recreation benefit studies," *Land Economics*, 52(1): 32–41.

Cho, S. H., Bowker, J. M. and Park, W. M., 2006, "Measuring the contribution of water and green space amenities to housing values: An application and comparison of spatially weighted hedonic models," *Journal of Agricultural and Resource Economics*, 31(3): 485–507.

Hanemann, W. M., 1978, *A Theoretical and Empirical Study of the Recreation Benefits from Improving Water Quality in the Boston Area*, Ph.D. Dissertation, Harvard University.

Hilger, J. and Hanemann, W. M., 2006, *The Impact of Water Quality on Southern California Beach Recreation: A Finite Mixture Model Approach*, Department of Agricultural and Resource Economics Working Paper, UC Berkeley.

Hoshino, T. and Kuriyama, K., 2010, "Measuring the benefits of neighbourhood park amenities: Application and comparison of spatial hedonic approaches," *Environmental and Resource Economics,* 45(3): 429–444.

Kelejian, H. H. and Robinson, D. P., 1993, "A suggested method of estimation for spatial interdependent models with autocorrelated errors, and an application to a county expenditure model," *Regional Science*, 72(3): 297–312.

——, 1995, "Spatial correlation: A suggested alternative to the autoregressive model," in L. Anselin and R. Florax (eds), *New Directions in Spatial Econometrics*, Berlin: Springer-Verlag, pp. 75–95.

——, 1997, "Infrastructure productivity estimation and its underlying econometric specifications: A sensitivity analysis," *Regional Science*, 76(1): 115–131.

Kim, C. W., Phipps, T. T. and Anselin, L., 2003, "Measuring the benefits of air quality improvement: A spatial hedonic approach," *Journal of Environmental Economics Management*, 45(1): 24–39.

Kim, J. and Goldsmith, P., 2009, "A spatial hedonic approach to assess the impact of swine production on residential property values," *Environmental and Resource Economics*, 42(4): 509–534.

Krinsky, I. and Robb, A. L., 1986, "On approximating the statistical properties of elasticities," *Review of Economics and Statistics*, 68(4): 715–719.

Kuriyama, K., Hanemann, W. M. and Hilger, J. R., 2010, "A latent segmentation approach to a Kuhn-Tucker model: An application to recreation demand," *Journal of Environmental Economics and Management*, 60(3): 209–220.

Leggett, C. G. and Bockstael, N. E., 2000, "Evidence of the effects of water quality on residential land prices," *Journal of Environmental Economics and Management*, 39(2): 121–144.

Paterson, W. R. and Boyle, K. J., 2002, "Out of sight, out of mind? Using GIS to incorporate visibility in hedonic property value models," *Land Economics*, 78(3): 417–425.

Phaneuf, D. J. and Smith, V. K., 2006, "Recreation demand models," in Mäler, K. G. and Vincent, J. R. (eds), *Handbook of Environmental Economics*, 1st edn, Volume 2, Elsevier, pp. 671–761.

Phaneuf, D. J., Kling, C. L. and Herriges, J. A., 2000, "Estimation and welfare calculations in a generalized corner solution model with an application to recreation demand," *Review of Economics and Statistics*, 82(1): 83–92.

Train, K. E., 2009, *Discrete Choice Analysis with Simulation*, 2nd edn, Cambridge, UK: Cambridge University Press.

von Haefen, R. H., 2003, "Incorporating observed choice into the construction of welfare measures from random utility models," *Journal of Environmental Economics and Management*, 45(2): 145–165.

von Haefen, R., and Phaneuf, D., 2005, "Kuhn–Tucker demand system approaches to non-market valuation," in R. Scarpa and A. Alberini (eds), *Applications of Simulation Methods in Environmental and Resource Economics*, Berlin: Springer, pp. 135–158.

von Haefen, R., Phaneuf, D. and Parsons, G. R., 2004, "Estimation and welfare analysis with large demand systems," *Journal of Business and Economic Statistics*, 22(2): 194–205.

Wales, T. J., and Woodland, A., 1983, "Estimation of consumer demand systems with binding non-negativity constraints," *Journal of Econometrics*, 21(3): 263–285.

4 Payment for agricultural ecosystem services and its valuation

Kentaro Yoshida

Introduction

Payments for ecosystem services (PES) have become a popular type of conservation scheme in the world. It is a useful tool to make people aware of the importance of biodiversity and ecosystem services, as well as their funding. Agriculture and forestry areas have been a major target for subsidies and grants from central and local governments in Japan. The governments implemented several environmental programs in order to preserve and enhance ecosystem services from agriculture and forests. Most of the environmental programs – for example, an environmental payment program, a forest environmental tax, and a rice terrace partnership program – have the characteristic of being PES or quasi-PES. In this chapter, the conservation of terraced paddy fields (rice terrace) and the related conservation programs, such as the direct payment program and the urban-rural partnership program, are selected for case studies of environmental valuation and its benefit transfer.

The scheme of direct payments for hilly and mountainous areas that started in the fiscal year 2000 is based on the idea of preserving the positive externalities of agriculture; that is, ecosystem services from agriculture. Compared with the environment payment policies being implemented in various countries, the Japanese direct payment system is lacking from the perspective of reducing environmental damage or risks caused by agriculture. With community agreements having been concluded in 26,937 locations in 1,723 cities, towns, or villages across the country in 2010, the establishment of an effective policy-evaluation method is required.

Under the direct payment policy, a community agreement is required to make it compulsory to preserve the positive externalities. However, the policy does not cover reduction of environmental damage. It does not call for actions that will necessitate changes in the ways of farming practice such as reduction in the use of fertilizers and agricultural chemicals. Therefore, the purpose of the Japanese direct payment scheme is different from that of the environmental payment scheme implemented in European Union nations. In this study, environmental valuation will be conducted from the aspect of environmental benefits, and also from the aspect of the environmental damage caused by the application of excessive amounts of agricultural chemicals and fertilizers. In order to see both the positive and negative environmental impacts, choice experiments can be adopted.

One of the advantages of the choice experiment method is that it allows researchers to evaluate and compare a number of different attributes using one question. If the Contingent Valuation Method (CVM) is used, only one scenario for these attributes can be questioned in any one questionnaire. With this advantage, choice experiments have recently received attention in environmental economics. However, as direct payments measures are conducted in a large number of small areas by individual local governments, budget constraints may make it difficult to apply a cost benefit analysis to the measures. Therefore, it appears to be important to assess the reliability of benefit transfers using choice experiments, although developing models with predictive power may be a formidable challenge (Haener *et al.* 2001).

Benefit transfer is a process that allows the prediction of a benefit estimate of a new policy site by using benefit estimates or functions from existing studies. If existing studies for a similar environmental good are applicable to the cost benefit analysis of the new policy site, government agencies will be able to avoid wasting time and money on repetitive work. It is often required by government agencies for practical reasons to obtain benefit estimates at a new policy site. Benefit transfer can be divided into three major types: fixed value transfer, expert judgment, and value estimator models (Bergstrom and De Civita 1999).

A number of empirical studies have been conducted to assess the transferability of benefit estimates. However, they have mostly focused on the contingent valuation method and the travel cost method (Downing and Ozuna 1996; Kirchhoff *et al.* 1997). Benefit transfer of choice experiments is a relatively new field, which should be explored through intensive empirical studies (Morrison and Bergland 2006).

In this study, after conducting choice experiments in almost identical forms in four locations – Kamogawa City, Koshoku City, Himi City, and Horai Town – the results of the experiments are examined to identify the transferability of benefit estimates. In the field of environmental and natural resource economics, research on benefit transfer and choice experiments has increased recently (Adamowicz *et al.* 1998; Bergstrom and De Civita 1999; Terawaki 2000; Kunimitsu *et al.* 2001; Yoshida 2000). However, it is relatively difficult to use choice experiments for the benefit transfer because the choice experiments elicit marginal willingness to pay (WTP) for each attribute. Marginal WTP depends on the attribute settings of each hypothetical scenario.

Since choice experiments have the advantage of producing a per unit benefit valuation of policy effects, it makes the connection with cost benefit analyses easy and it is efficient in terms of cost and time. If the benefit-transfer approach of choice experiments is established, it will further enhance the efficiency of cost benefit analyses of environmental programs.

Environmental valuation by choice experiments

Stated preference methods

Environmental valuation is divided into two categories: revealed preference and stated preference. Revealed preference – like the travel cost method and hedonic price method – is a method to reveal the environmental value reflected in existing market data, such as travel costs and land prices. Stated preference, as represented by CVM, is a method to reveal environmental value based on the valuations stated by beneficiaries in a survey questionnaire.

Research on environmental valuation has become popular in Japan since the 1990s, as it is in many other countries. CVM has played a key role in such research, as environmental valuation methods have been increasingly adopted for cost benefit analyses of environmental policy and public works. However, CVM has its limitations. It can evaluate only a combination of single attributes or levels per question. For example, when evaluating the ecosystem services of agriculture, CVM allows the examination of only one kind of scenario, such as implementing a policy to prevent a decline of the ecosystem services of a region by 30 percent over the next 10 years. Therefore, in order to compare different kinds of policy alternatives, it is necessary to develop several different questionnaires. Ecosystem services can be classified into a variety of services, such as landscape conservation and national land conservation. However, it is difficult to calculate the value of each service by CVM. On the other hand, conjoint analysis not only allows direct comparison of various policy alternatives by using one questionnaire, but it also reveals the valuation of each service.

When conducting a cost benefit analysis of environmental policies, it is necessary to evaluate the benefit in monetary terms. If CVM is applied in such a case, budgetary and time constraints often become a problem. The application of benefit transfer has been studied. However, in the case where environmental policies are implemented separately in various regions of the country, the extent of the policy effects will be different from one region to another. In such a case, per unit benefit estimates of policy effects become necessary. It is often difficult to estimate a basic unit by CVM. However, the conjoint analysis produces benefit valuations in the form of marginal WTP which allows using it as the basic unit for the cost benefit analysis.

There are various types of conjoint analysis and each has its advantages and disadvantages. Since this study is aimed at investigating whether a policy should be implemented or not, and also at comparing policy alternatives, choice experiments are adopted for the valuation of ecosystem services from agriculture.

Data collection

Data were obtained by mail surveys. Questionnaires were sent to general households selected with a random sampling method, using telephone directories, in the study's three cities and one town. The survey in Kamogawa City was conducted in July 2001 and those in the other three locations in November 2001. The response rate was 82 percent in Kamogawa City (164/200), 87.3 percent in Koshoku City (172/197), 82.9 percent in Himi City (165/199), and 90.0 percent in Horai Town (180/200). By sending a reminder twice using the same method as adopted by Mangione (1995), we achieved a sufficient response rate for a mail survey.

Hypotheses and empirical methods

Benefit transfer is a method to estimate benefit valuations by using existing studies for the areas which are covered by a new policy and require benefit evaluation. With regard to the direct payment scheme, if it is possible to obtain the benefit valuations of all municipalities covered by the policy from the benefit evaluations conducted in several scores of municipalities, it is very efficient in terms of budgetary and time constraints.

In this study, the transferability of the results of the benefit evaluations conducted in the four municipalities was examined. The transferability was tested by a likelihood ratio test – which incorporates a relative scale factor – from the perspective of benefit-function transfer (Ben-Akiva and Lerman 1985; Swait and Louviere 1993). In addition to it, a complete combinatorial approach was used for the transferability test of benefit estimates (Poe *et al.* 1994; Poe *et al.* 2005). The non-overlapping confidence interval criteria might be a possible option for the transferability test, but it was less robust for the comparison of benefit estimates. If it passes both tests, it can be said that the transferability of benefit functions has been guaranteed.

The first is the test of log-likelihood of each benefit function. Suppose there are two samples, assuming that the specifications of the models are identical. The following hypothesis is tested.

$$H_1 : \beta_1 = \beta_2 \quad \text{and} \quad \mu_1 = \mu_2$$

where μ_1 and μ_2 are two scale factors, and β_1 and β_2 are corresponding parameter vectors. At first, the equality of β_1 and β_2 is tested.

$$H_{1A} : \beta_1 = \beta_2 = \beta,$$

while permitting the scale factors to differ between data sets. If H_{1A} is rejected, H_1 is also rejected. If H_{1A} cannot be rejected, then test the hypothesis:

$$H_{1B} : \mu_1 = \mu_2 = \mu.$$

Both H_{1A} and H_{1B} can be tested using standard likelihood ratio test statistics.

On the other hand, complete combinatorial is used for the transferability test of benefit estimates (MWTP). Complete combinatorial is to calculate every possible difference between the two empirical distributions of the benefit estimates $MWTP_1$ and $MWTP_2$. If the null hypothesis is not rejected, the transferability of benefit estimates from two samples is considered to be guaranteed:

$$H_2 : MWTP_1 - MWTP_2 = 0.$$

Outline of choice experiments

One of the advantages of conjoint analysis as a tool for environmental valuation is that it can be formularized on the basis of random utility theory. Since the choice experiment is in the form of selecting one alternative from among several kinds of policy alternatives, it can be formularized as follows.

When the ith respondent chooses j from among J number of options, utility (U) is expressed as follows.

$$U_{ij} = V_{ij} + \varepsilon_{ij}$$

V is the deterministic component of utility and ε is a stochastic component of utility. Therefore, when respondent i chooses j, the utility is higher than in the case of choosing other options. It can be formularized as follows.

$$\begin{aligned} \text{Prob}\,(j) &= \text{Prob}\,(U_{ij} > U_{ik}; \forall k \neq j) \\ &= \text{Prob}\,(V_{ij} + \varepsilon_{ij} > V_{ik} + \varepsilon_{ik}; \forall k \neq j) \\ &= \text{Prob}\,(V_{ij} - V_{ik} > \varepsilon_{ik} - \varepsilon_{ij}; \forall k \neq j) \end{aligned}$$

McFadden (1973) showed that as long as J stochastic components follow the type I extreme value distribution, the probability of choosing option j can be expressed as follows:

$$\text{Prob}\,(j) = \exp(\beta x_{ij}) / \sum \exp(\beta x_{ik}).$$

This is a conditional logit model. Here, we study a main effect model that limits the deterministic utility function V to attribute x_{ij} which is peculiar to the options. Models which incorporate taste variances of respondents, for example a random parameters logit model, or a latent class model, have been used more frequently recently. However, a classic conditional logit model facilitates the demonstration of the accuracy of each test and their comparison.

Design of choice experiments

A profile design is important in conducting choice experiments. In other words, what attributes and levels should be presented as policy alternatives to respondents? With regard to environmental benefits that can be maintained and conserved by the direct payment scheme for hilly and mountainous areas, we selected as attributes the ecosystem services concerning rural amenities, and the ecosystem services concerning national land conservation. Specifically, the following explanations were provided: roles to conserve the landscape of rice terraces and the environment for living creatures, such as aquatic insects, dragonflies, and fireflies (conservation of rural landscape and environment for living creatures); and roles to prevent disasters, such as floods and landslides (protection against disasters and national land conservation). For the environmental damage that is the negative impact of the policies, we selected water pollution. Specifically, the following explanation was provided: the effects of agricultural chemicals and fertilizers that pollute rivers and groundwater, affecting people's lives and damaging the environment for living creatures (river and groundwater contamination). In addition, we incorporated a contribution to funds and determined each attribute and level. We set four levels for the two attributes of environmental benefits from −30 percent to +60 percent; four levels for water pollution from −50 percent to the status quo; and six levels for the contribution to the fund from 0 to 10,000 yen.

As a hypothetical situation in choice experiments, in our original survey, we provided the following explanation.

Suppose we establish a rice terrace preservation fund in order to protect the rice terraces in the city and collect contributions from the people living in the city or the prefecture to raise funds to implement various measures. For example, we use the funds to purchase rice terraces that are extremely important in terms of ecosystem services, or we use the funds to invite volunteers to restore abandoned rice terraces.

Another possible measure is to stock rice paddies with fireflies and aquatic insects. Through such measures, the positive impacts of rice terraces on the environment will be maintained and enhanced. However, an increase in rice production in rice terraces means an increase in the use of agricultural chemicals and fertilizers. This may result in aggravating the negative impact on the environment of rice terraces.

After this explanation, we presented options to the respondents five times and asked them to select only one option for each measure (see Table 4.1). We determined combinations of profiles on the basis of the orthogonal factorial design. We excluded unrealistic combinations whose conditions are worse than in Measure 4, such as those demanding contributions despite the fact that all attributes deteriorate. Measure 4 is common to all choice sets. It has been confirmed that the evaluation of environmental protection policies by choice experiments is influenced by status quo bias. This is because respondents believe that they are endowed with the current state of the environment. It is also called the endowment effect (Adamowicz *et al.* 1998; Garrod and Willis 1999).

Results

Table 4.2 shows that each parameter becomes significantly different from 0 at the significance level of 1 percent. Table 4.3 shows that the marginal WTP concerning

Table 4.1 Examples of profiles

	Measure 1	*Measure 2*	*Measure 3*	*Measure 4*
Water contamination (%)	−30	−10	−30	status quo
National land conservation (%)	+60	+60	status quo	−30
Rural amenities (%)	+30	status quo	status quo	−30
Contribution to fund	5,000 yen	1,000 yen	500 yen	0 yen

Table 4.2 Estimated results from a conditional logit model

Variable	Kamogawa	Koshoku	Himi	Horai
Water contamination (%)	−0.02440	−0.04515	−0.03330	−0.03716
	(−7.287)	(−12.490)	(−10.057)	(−10.468)
National land conservation (%)	0.005746	0.01206	0.01028	0.009880
	(2.845)	(6.230)	(5.158)	(5.075)
Rural amenities (%)	0.01031	0.01301	0.01255	0.009648
	(5.355)	(7.003)	(6.555)	(5.133)
Contribution to fund (JPY)	−0.0001969	−0.0002209	−0.0002227	−0.0002170
	(−8.042)	(−9.687)	(−9.326)	(−9.181)
Alternative-specific constant	−1.078	−0.9745	−1.085	−0.8990
	(−4.912)	(−4.743)	(−5.033)	(−4.34)
Number of observations	543	680	614	620
Adjusted R-squared	0.1263	0.2063	0.1826	0.1583

Note: Figures in the parentheses are asymptotic t value; JPY is Japanese yen.

Table 4.3 Estimated results of marginal willingness to pay (Japanese yen)

	Kamogawa City	Koshoku City	Himi City	Horai Town
Water contamination	−123.9 [−177.4, −85.4]	−204.4 [−261.3, −162.7]	−149.5 [−196.9, −114.8]	−171.2 [−221.8, −135.3]
National land conservation	29.2 [10.2, 49.0]	54.6 [37.5, 74.6]	46.1 [29.2, 65.7]	45.5 [28.6, 65.4]
Rural amenities	52.3 [32.7, 76.5]	58.9 [41.1, 80.8]	56.3 [37.8, 78.6]	44.5 [25.4, 66.4]

Note: Figures in parentheses are lower and upper bounds of the 95% confidence interval.

the positive environmental effects (landscape and environment for living creatures; rural amenities; and disaster prevention and national land conservation) in only Kamogawa City is significantly different statistically from other cities, and that there is no significant difference in the other models. It also shows that the absolute value of the marginal WTP for water contamination is about three times as large as it is for environmental benefits.

Table 4.4 shows the results of both the likelihood ratio tests on benefit function transfer and the complete combinatorial on benefit estimates transfer. As to the likelihood ratio tests, none of the null hypotheses for the combinations were rejected except for those with Kamogawa-Koshoku, and Kamogawa-Horai, and the transferability of benefits was confirmed. As to the complete combinatorial, most of the transferability tests were not rejected at a significance level of 5 percent. The only three attributes of two combinations were: Kamogawa-Koshoku

Table 4.4 Transferability tests of benefit function and benefit estimates

	Likelihood ratio test	Complete combinatorial		
		Water contamination (%)	National land (%)	Rural amenities (%)
Kamogawa–Koshoku	24.30**	0.7%**	2.8*	32.8
Kamogawa–Horai	3.88*	20.8	10.6	38.0
Kamogawa–Himi	7.20	5.6	10.6	30.3
Koshoku–Horai	0.00	4.1*	25.7	42.2
Koshoku–Himi	0.00	16.3	24.3	15.3
Horai–Himi	0.00	23.6	46.4	22.0

Note: ** and * denote rejection at a significance level of 1% and 5%, respectively.

(water contamination, and national land conservation), and Koshoku-Horai (water contamination).

Conclusions

In this chapter, benefit transfer by choice experiments was tested using the results of surveys conducted in four municipalities. Four municipalities have similar types of rice terrace resources and they participate in the direct payment scheme. As a basis for implementing such a quasi-PES program to conserve rural amenities, benefit valuation and benefit transfer clearly demonstrate the importance of public and private support. Benefit estimates show only a small difference between different policy sites. It continues to be a signal and driving force for the implementation of PES.

Yoshida (2000) conducted benefit transfer by CVM and found that with regard to combinations whose benefit function transfer is possible, direct transfer of benefit estimates is also possible. The benefit transfer by choice experiments confirmed a similar trend, although the test used for this study is slightly different from the study using CVM. By conducting both likelihood ratio tests and complete combinatorial, it was found that at least two combinations out of six possible combinations were transferable to each other. The only exception was a function estimated from the data for Kamogawa City. It may be significant that only in Kamogawa City was the survey conducted when rice was actually growing in rice terraces. The surveys in the three other municipalities were conducted after rice was reaped.

The results of the analyses may indicate that evaluation reflects the conditions of the landscape and the environment at the time of the survey. For example, the evaluation of national land conservation and amenity in Kamogawa City shows a significant difference, while evaluations in the three other municipalities show no significant difference. Evaluation by choice experiments is likely to be influenced by price fluctuations, depending upon when a survey is conducted. Therefore, when conducting benefit transfer, it is necessary to pay special attention to this effect. This might be done by conducting the surveys at the same time.

References

Adamowicz, V., Boxall, P., Williams, M. and Louviere, J., 1998, "Stated preference approach for measuring passive use values: Choice experiments and contingent valuation," *American Journal of Agricultural Economics*, 80(1): 74–75.

Ben-Akiva, M. and Lerman, S. R., 1985, *Discrete Choice Analysis: Theory and Application to Travel Demand*, Cambridge, MA, MIT Press.

Bergstrom, J. C., and De Civita, P., 1999, "Status of benefits transfer in the United States and Canada: A review," *Canadian Journal of Agricultural Economics*, 47: 79–87.

Downing, M. and Ozuna, T. Jr., 1996, "Testing the reliability of the benefit function transfer approach," *Journal of Environmental Economics and Management*, 30(3): 316–322.

Garrod, G. and Willis, K. G., 1999, *Economic Valuation of the Environment: Models and Case Studies*, Cheltenham, Edward Elgar.

Haener, M. K. Boxall, P. C. and Adamowicz, W. L., 2001, "Modeling recreation site choice: Do hypothetical choices reflect actual behavior?" *American Journal of Agricultural Economics*, 83(3): 629–642.

Kirchhoff, S., Colby, B. G. and LaFrance, J. T., 1997, "Evaluating the performance of benefit transfer: An empirical inquiry," *Journal of Environmental Economics and Management*, 33(1): 75–93.

Kunimitsu, Y., Matsuo, Y. and Tomosyo, T., 2001, "The causative factors about contingent valuation of rural park and the transferability of WTP function: Mainly about individual attributes, consolidation conditions, and area situations," *Journal of Rural Planning Association*, 20(1): 31–40.

Louviere, J. J., Hensher, D. A. and Swait, J. D., 2000, *Stated Choice Methods: Analysis and Application*, Cambridge, Cambridge University Press.

McFadden, D., 1973, "Conditional logit analysis of qualitative choice behavior", in P. Zarembka (ed.) *Frontiera in Econometrics*, New York, Academic Press.

Mangione, T. W., 1995, *Mail Surveys: Improving the Quality*, London, Sage Publications.

Morrison, M. and Bergland, O., 2006, "Prospects for the use of choice modeling for benefit transfer," *Ecological Economics*, 60(2): 420–428.

Poe, G. L., Severance-Lossin, E. K. and Welsh, M. P., 1994, "Measuring the difference $(X - Y)$ of simulated distributions: A convolutions approach," *American Journal of Agricultural Economics*, 76(4): 904–915.

Poe, G. L., Giraud, K. L. and Loomis, J. B., 2005, "Computational methods for measuring the difference of empirical distributions," *American Journal of Agricultural Economics*, 87(2): 353–365.

Swait, J. and Louviere, J., 1993, "The role of the scale parameter in the estimation and use of multinomial logit models," *Journal of Marketing Research*, 30(3): 305–314.

Terawaki, T., 2000, "Benefit function transfer in agricultural public work," *Journal of Rural Economics*, 71(4): 179–186.

Tsuge, T., 2001, "The valuation of the externalities of forest on citizen's preferences and possibility of use in a policy-making: A choice experiment study," *Environmental Science*, 14(5): 465–476.

Yoshida, K., 2000, "Assessing the convergent validity of environmental benefit estimates by a benefit transfer approach: Applications of meta-analysis and benefit function transfer," *Journal of Rural Economics*, 72(3): 122–130.

5 Determinants of happiness

Environmental degradation and attachment to nature

Tetsuya Tsurumi, Kei Kuramashi, and Shunsuke Managi

Introduction

Most countries have set economic growth as one of their highest priority objectives. Survey evidence, however, suggests that the relationship between income (GDP per capita) and the level of individual self-reported happiness (subjective well-being index) is not linear. Figure 5.1 presents a simple scatter plot of income and happiness for 94 countries from 1952 to 2004. Although there is an increasing trend between happiness and income, we find that after a certain level of income, the marginal utility of income decreases.

As represented by Maslow's hierarchy of needs (Maslow 1954), there is a possibility that people's happiness increases more when their basic human needs (or deficiency needs) are fulfilled, rather than when their additional or diversified needs are fulfilled. Because basic human needs tend to be satisfied as people obtain a minimum income, a minimum income seems to be strongly related to happiness. Economists also discuss the relationship between income and happiness (Nordhaus and Tobin 1972; Easterlin 1974; Scitovsky 1976).[1] Their basic argument is that income and economic growth comprise only a small portion of the equation that makes people happy, because an increase in income leads to other related problems such as an increase in working hours or environmental degradation, which seem to have a negative correlation with happiness.

Over the past decades, studies concerning happiness have identified a number of variables that seem to explain happiness. Following Frey and Stutzer (2001), we consider four factors that may affect people's subjective well-being: economic, socio-demographic, personality, and environmental.

Besides income, among the economic factors unemployment is thought to be one of the most important. For example, Tella *et al.* (2001) find that a sudden increase in the unemployment rate by 1 percent results in an approximately 10 percent drop in individuals' self-reported well-being levels. Many studies take into consideration socio-demographic factors such as age, sex, marital status, and health. Peiro (2006) finds that happiness with respect to age typically presents in a parabolic shape, reaching its minimum at about the age of 40 years, that health has

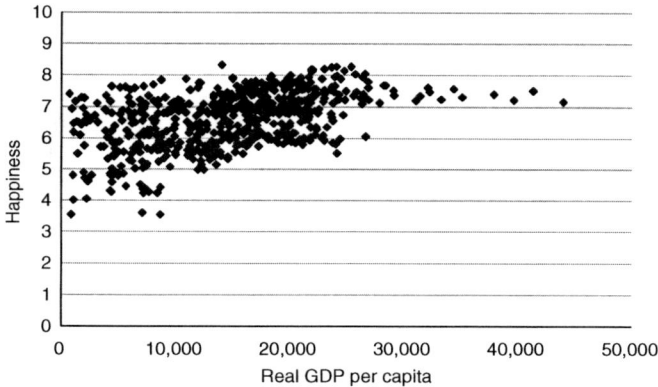

Figure 5.1 Simple scatter plot between happiness and real GDP per capita.

a strong relationship with happiness, and that married people are usually happier than others. With regard to personality factors, there is strong evidence that about 50 percent of the variation in the measures of subjective well-being are due to inherited personality traits (Layard 2011). We therefore control for personality factors in this study.

In this chapter, our main objective is to clarify whether environmental degradation and attachment to nature affect subjective well-being. If attachment to nature has a positive correlation with happiness, we can say environmental education or increasing the time we take to be in touch with nature have the possibility of increasing our happiness. A potential problem is that positive or negative attitudes toward the environment may change owing to underlying psychological characteristics, economic factors such as income and unemployment, and socio-demographic factors such as age, sex, marital status, and health. We therefore control for these potential factors too.

The remainder of this chapter is organized as follows. First, in the next section, we examine the effect of environmental degradation on happiness, using global-level data. In the third section, we use country-level data obtained from our original survey in Japan to test the relationship between environmental degradation and happiness. In the section that follows, we examine the effect on happiness of attachment to nature. In the final section we present our findings.

Environmental degradation and happiness: Global-level analysis

Literature

The relationship between environmental degradation and happiness is a relatively new area of research, with only a few studies in this field. Welsch (2002) uses cross-sectional data from 54 countries to examine the relationship between subjective well-being and pollution from pollutants such as sulfur dioxide (SO_2), phosphorus, suspended solids, and nitrogen dioxide (NO_2). He obtains a statistically significant negative effect on subjective well-being only for NO_2. To avoid unobserved heterogeneity and the danger of spurious correlation, Welsch (2006) employs a panel data approach. Using the data on NO_2, particulate matter (PM10), and lead, for ten countries, from 1990 to 1997, he employs fixed effect estimation to obtain a statistically significant negative effect only for NO_2. Welsch (2007) employs seemingly unrelated regression (SUR), using cross-country data for NO_2, and obtains statistically significant results.

We note that Welsch (2002; 2006; 2007) uses self-reported well-being data from the World Database of Happiness (WDH). Derived from surveys, the WDH measures the average level of well-being in a given country. The surveys are based on the following question: Taking all things together, would you say you are very happy, quite happy, not very happy, or not at all happy? The responses are rated from 1 (not at all happy) to 4 (very happy). This dataset consists of 58 countries, and refers to the early and mid 1990s.

Data and estimation results

We employ both parametric regression and semi-parametric regression to relax the assumption of linearity.

Parametric approach

Our model is as follows:

$$\ln H_{it} = \alpha_{1i} + \gamma_{1t} + \beta_1 \ln GDP_{it} + \beta_2 \ln UNEMPLOYMENT_{it} + \beta_3 \ln E_{it}^k + \varepsilon_{1it}.$$

Here, i denotes the country, t denotes the year, H_{it} stands for subjective well-being, GDP_{it} is real gross domestic product per capita, $UNEMPLOYMENT_{it}$ denotes the unemployment rate, E_{it}^k denotes the k-type environmental index (k = PM10, SO_2, CO_2, and energy use), α_{1i} is a country-specific effect, γ_{1t} is a time-specific effect, and ε_{1it} is the error term. We separate our specifications by each environmental index to avoid multicollinearity.

We draw data for this study from several sources. H_{it} comes from the World Value Survey. We adopt the use of happiness data from the World Value Survey because in terms of sample observations, this dataset is superior to that obtainable from the WDH. The scores are based on responses to the question: All things considered, how satisfied or dissatisfied are you with your life as a whole now? The scores range from 0 (dissatisfied) to 10 (satisfied).

With regard to data for environmental indices, we use PM10 and SO_2 as local pollutants that have already been tested in previous studies, and CO_2 and energy use as global indices that have never been tested.

Data on PM10 come from the World Development Indicators. PM10 refers to the fine suspended particulates less than 10 microns in diameter, capable of penetrating deep into the respiratory tract and causing significant damage to health.

The estimates represent the average annual exposure level of the average urban resident to outdoor particulate matter. The state of a country's technology and pollution control is an important determinant of particulate matter concentrations. The Center for Air Pollution Impact and Trend Analysis (CAPITA) and Stern (2005) produced a comprehensive database on SO_2 emissions. Their estimates are considered superior to others in terms of spatial and temporal resolution and extent (see Stern 2005). SO_2 is produced mainly from the combustion of petroleum and coal, and it is the key pollutant causing three important environmental problems: local air pollution and smog, acid rain, and dry deposition. It exerts a harmful influence on the respiratory system. CO_2 emissions stem from the burning of fossil fuels and the manufacture of cement. This includes CO_2 produced during the burning of solid, liquid, and gas fuels and gas flaring.[2] While SO_2 has local and trans-boundary impacts, CO_2 is a greenhouse gas with a global impact. See Cole and Elliott (2003) for a more comprehensive comparison of these emissions. Energy use refers to the use of primary energy such as coal and coal products, oil and petroleum products, natural gas, and nuclear and hydroelectric power before their transformation into other end-use fuels.[3] While CO_2 emissions stem from the burning of fossil fuels and the manufacture of cement, it is notable that fossil fuels are the main source of CO_2 emissions. The difference between the data on CO_2 emissions and energy use is that the latter take into consideration not only fossil fuels but also the primary energy sources that do not emit CO_2, such as nuclear and hydroelectric power. This difference is important when we interpret the estimation results in the next section. CO_2 and energy use data are obtained from the World Development Indicators.

GDP_{it}, which is defined in constant 2000 US dollars, is obtained from Penn World Table 6.1, and $UNEMPLOYMENT_{it}$ and population data are obtained from the World Development Indicators. Our panel data cover 50 countries from 1980 to 2000. Table 5.1 lists our sample countries.

Table 5.2 shows the estimation results. We show the results of a fixed effects model, and not a random effects model, because the Hausman test indicates that the random effects model may not be consistent. Although we obtain the expected signs with regard to income and unemployment, we do not find statistically significant results concerning the environmental degradation indices. These results are consistent with Welsch (2002; 2006; 2007), but there is a possibility that when we employ parametric estimation, our functional assumption concerning the environmental degradation indices may be too strong. To deal with this potential problem, we employ the semi-parametric approach detailed below.

Semi-parametric approach

Semi-parametric regression analysis relaxes the assumption of linearity and typically substitutes the weaker assumption that the average value of a response is a smooth function of the predictors.

Although several semi-parametric and nonparametric methods can be used to estimate the regression line, there are usually two obstacles. First, as the number

Table 5.1 List of countries

Europe	Africa	Latin America
Iceland	Egypt	Argentine
Ireland	Nigeria	Uruguay
Italy	South Africa	Ecuador
Austria	**Middle East**	El Salvador
Netherland	Israel	Guatemala
Greece	Iran	Costa Rica
Switzerland	**Asia**	Columbia
Sweden	Pakistan	Chile
Spain	Bangladesh	Dominica
Denmark	Philippines	Nicaragua
Hungary	Hong Kong	Panama
Finland	North Korea	Paraguay
France	Japan	Brazil
Belgium	**Oceania**	Venezuela
Poland	Australia	Peru
Portugal	**North America**	Honduras
Rumania	America	Mexico
Luxembourg	Canada	
United Kingdom		

Table 5.2 Estimation results (fixed effects)

ln(Happiness)	Model (a)	Model (b)	Model (c)	Model (d)
ln(GDP)	0.135**	0.171***	0.158***	0.156***
	(2.12)	(4.46)	(3.86)	(4.04)
ln(UNEMPLOYMENT)	−0.009	−0.020*	−0.020*	−0.020*
	(−0.62)	(−1.82)	(−1.86)	(−1.87)
ln(PM10)	−0.080			
	(−1.32)			
ln(SO$_2$)		−0.013		
		(−1.30)		
ln(ENERGYUSE)			−0.003	
			(−0.09)	
ln(CO$_2$)				−0.002
				(−0.08)
R^2	0.172	0.227	0.222	0.222
Observations	222	344	344	344
Number of id	49	50	50	50

Note: Values in parentheses are *t*-values. *, ** and *** denote significance at the 10%, the 5%, and 1% levels, respectively.

of independent variables increases, the sparseness of data inflates the variance of estimates. This problem of rapidly increasing variance is called the curse of dimensionality. Second, because most of the semi-parametric and nonparametric regressions do not provide an equation relating the average response to the independent variables, the results become difficult to interpret[4] (see Hastie and Tibshirani 1990).

These problems led to the development of the additive regression models, which have two advantages (Stone 1985). First, because each of the individual additive terms is estimated using a univariate smoother, the curse of dimensionality is avoided. Second, the interpretation of additive models is relatively simple because a two-dimensional plot is sufficient to examine the estimated partial regression function, holding the other independent variables constant.

In this study, we use generalized additive models (Hastie and Tibshirani 1990). We use cubic spline smoothing[5] iteratively to minimize the partial residuals that remain after removing the influence of the other variables in the model. In this iteration, the estimation loop stops when the model fit cannot be improved.

Our estimation model is as follows:

$$\ln H_{it} = \rho_2 + \alpha_{2i} + \gamma_{2t} + f_1(GDP_{it}) + f_2(UNEMPLOYMENT_{it}) + f_3(E_{it}^K) + \varepsilon_{2it},$$

where ρ_2 is a constant term and $f(\cdot)$ is a generic flexible functional form allowing for a potentially nonlinear non-monotonic relationship.[6]

The resulting scatter plots in Figures 5.2 through 5.7 show the predicted contributions to the dependent variable from the independent variables. First, Figure 5.2 shows the predicted contribution to happiness from income. The central curve represents the estimated results. The upper and lower curves correspond to the upper and lower 95 percent confidence intervals. Figure 5.2 shows an increasing trend at all income levels, implying that after removing the other independent variables' effects, the relationship between income and happiness would show a monotonic increasing trend. This result is consistent with our parametric regression. It should be noted that we do not find the marginal utility of income to be decreasing. Kahneman and Deaton (2010) suggest that a high income improves the evaluation of life, but not of emotional well-being. Their term emotional well-being refers to the emotional quality of an individual's everyday experience. On the other hand, life evaluation refers to the thoughts that people have about their life when they think about it explicitly. The happiness index we use seems to be of a life-evaluation type, which may result in our monotonic increasing trend in Figure 5.2.

Second, Figure 5.3 shows the relationship between the unemployment rate and happiness. We find a slightly decreasing trend between them, suggesting that an increase in the unemployment rate results in a decrease in happiness, as expected.

Figures 5.4 through 5.7 present our main concern – the relationship between environmental degradation and happiness. First, Figure 5.4 shows the results of PM10, in which we find an apparently decreasing trend. This may be because PM10 is easy to perceive and it causes significant health problems in many places.

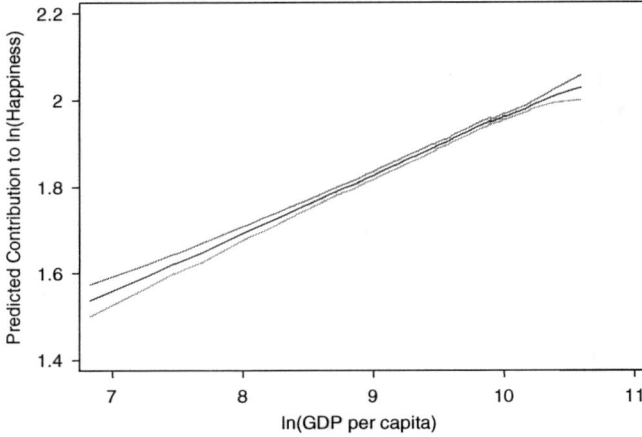

Figure 5.2 Happiness and real GDP per capita.

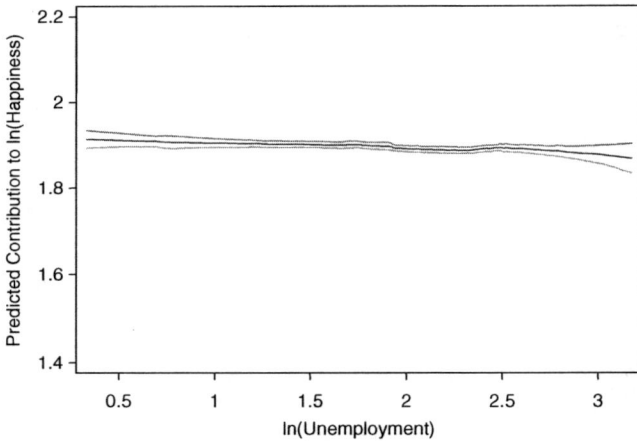

Figure 5.3 Happiness and unemployment.

Figure 5.5 depicts SO_2 emissions. While we find a flat tendency at the lower and middle emission levels, at the high emission level we find a slightly decreasing trend. This implies that when the SO_2 emission level exceeds this threshold point, it can lead to serious problems such as health damage or acid rain.

On the other hand, as shown in Figures 5.6 and 5.7, an increase in the CO_2 emission level and energy use do not affect one's subjective well-being. This implies that people do not have much concern about these global indices. However, it should be noted that if people increase their concerns about global warming or

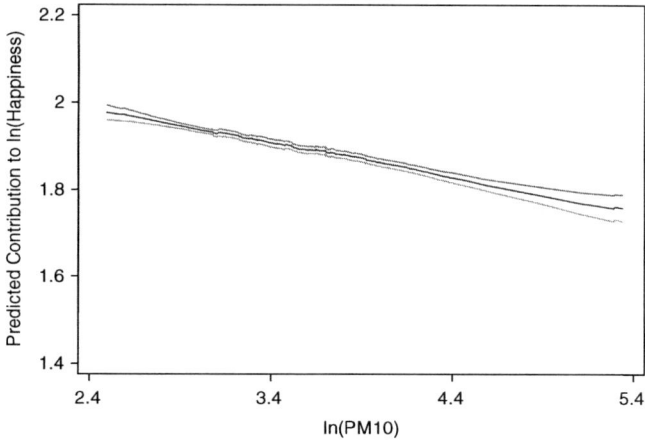

Figure 5.4 Happiness and PM10 concentration.

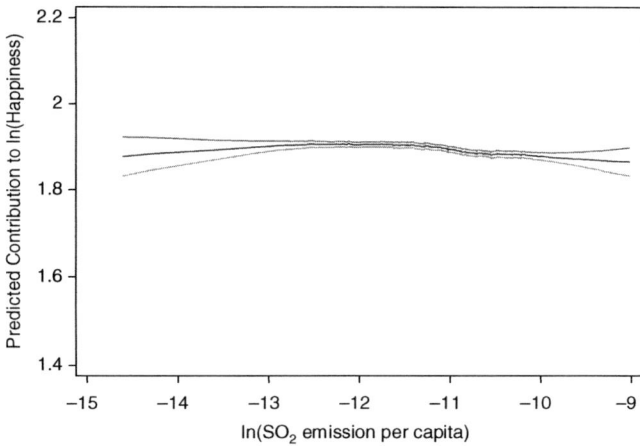

Figure 5.5 Happiness and per capita SO_2 emissions.

resource depletion in the future, an increase in CO_2 emissions or energy use may affect their subjective well-being.

Environmental degradation and happiness: Country-level analysis

Environmental degradation and happiness

In this subsection, we employ individual-level estimation, with data from a survey conducted in Japan. As in the previous section, our objective is to identify the

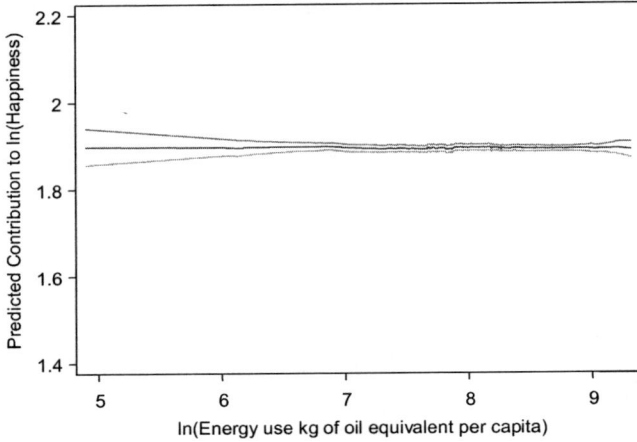

Figure 5.6 Happiness and per capita energy use.

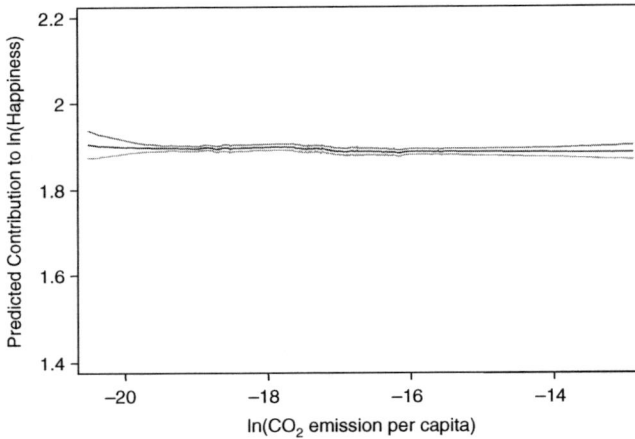

Figure 5.7 Happiness and per capita CO_2 emissions.

relationship between environmental degradation and happiness after controlling for certain potential factors for happiness such as economic, socio-demographic, and personality factors.

We use the following environmental indices: SO_2 concentration, NO_2 concentration, carbon monoxide (CO) concentration, photochemical oxidant (Ox) concentration, and suspended particulate matter (SPM) concentration. The survey was conducted in the Tokyo and Kanagawa prefectures in Japan in January 2009.

We obtained the environmental indices from the local monitoring data presented by the Atmospheric Environmental Regional Observation System (AEROS). The concentration data are monitored in many local points in Japan. We merge these data with our individual survey data according to the place where the respondents live. Because local pollutant concentrations are different even within the same prefecture, our estimations using the local area data have an advantage compared with country-level estimation. In addition, this study is the first to examine the relationship between environmental degradation and happiness in Japan using individual survey data.

Model and data

As in the previous section, we employ both parametric estimation and semi-parametric estimation to consider the potential problems of our assumption of the functional form.

Parametric estimation

In our analysis, in order to avoid multicollinearity, we divide our specifications by environmental indices. Our estimation model is as follows:

$$H_i = \rho_3 + \delta_1 INCOME_i + \delta_2 UNEMPLOYMENT_i + \delta_3 HEALTH_i$$
$$+ \delta_4 SHOCK_i + \delta_5 SEX_i + \delta_6 TIMEDISCOUNT_i$$
$$+ \delta_7 ANXIETY_i + \delta_8 ALTRUISM_i + \delta_9 E_i^K + \varepsilon_{3i}$$

Here, i denotes an individual and H_i denotes the individual's subjective well-being index based on responses to the question: All things considered, how satisfied or dissatisfied are you with your life as a whole now? The scores range from 0 (dissatisfied) to 10 (satisfied). $INCOME_i$ is the annual household income and $UNEMPLOYMENT_i$ is the dummy variable concerning unemployment unrest that takes a value of 1 if the individual feels unrest and 0 otherwise; $HEALTH_i$ denotes the health unrest dummy that takes a value of 1 if the individual feels unrest and 0 otherwise; $SHOCK_i$ denotes the number of shocking experiences; SEX_i is the sex dummy that takes a value of 1 if the person is male and 0 otherwise; $TIMEDISCOUNT_i$ is a proxy for time discounting that takes a large value if a person has a large time-discount rate; $ANXIETY_i$ denotes the risk-averse attitude dummy that takes a large value if a person is risk-averse; $ALTRUISM_i$ denotes the altruism dummy that takes a value of 1 if a person is altruistic and 0 otherwise; and E_i^k denotes a k-type environmental index ($k = SO_2$, NO_2, CO, Ox, and SPM). The term ρ_3 is a constant and ε_{3i} is an error term. Here, we separate our specifications by each environmental index to avoid multicollinearity. With regard to the data for environmental indices, we use both the mean concentration during the survey period (from 20 to 22 January) and the maximum concentration,

Table 5.3 Survey outline

Period	20 January to 22 January 2009
Target	1,043 households in Tokyo and Kanagawa
Method	Internet survey
Survey item	(1) household income (INCOME)
	(2) unemployment unrest (UNEMPLOYMENT)
	(3) health unrest (HEALTH)
	(4) number of shocking events (SHOCK)
	(5) sex (SEX)
	(6) time discounting (TIMEDISCOUNT)
	(7) risk aversion (ANXIETY)
	(8) altruism (ALTRUISM)
	(9) subjective well-being (HAPPINESS)

Table 5.4 Descriptive statistics

Variable	Average	SD	Min	Max	Obs	Unit
SO_2 (MEAN)	0.00322	0.00120	0.0000769	0.00745	796	ppm
SO_2 (MAX)	0.00666	0.00416	0.00100	0.0170	796	ppm
NO_2 (MEAN)	0.0319	0.00451	0.0135	0.0475	865	ppm
NO_2 (MAX)	0.0485	0.00648	0.0230	0.0880	865	ppm
CO (MEAN)	0.563	0.174	0.307	0.989	190	ppm
CO (MAX)	1.13	0.508	0.400	2.20	190	ppm
Ox (MEAN)	0.00843	0.00460	0.00166	0.0217	758	ppm
Ox (MAX)	0.0223	0.00657	0.00500	0.0460	758	ppm
SPM (MEAN)	0.0273	0.00710	0.00933	0.0550	879	mg/m^3
SPM (MAX)	0.0511	0.0139	0.0110	0.102	879	mg/m^3
INCOME	602	323	50	1450	999	10,000 yen
SHOCK	1.62	2.64	0	21	999	Number
TIMEDISCOUNT	50.5	50.4	0	999	999	Minutes
ANXIETY	6.88	1.97	0	10	999	–

because there is a possibility that only maximum concentrations relate to happiness. We present the survey outline and descriptive statistics in Tables 5.3 and 5.4, respectively.

Tables 5.5 and 5.6 show the estimation results; we find the expected signs concerning psychological characteristics, economic factors, and socio-demographic factors in both tables. On the other hand, we do not find statistically significant results with regard to environmental degradation indices, with the exception of the maximum Ox concentration (Model (i)), which suggests that the degradation of maximum Ox concentration results in a decrease in happiness. To address the potential problems of the functional form, we discuss semi-parametric estimation in the next subsection.

Table 5.5 Estimation results: mean concentration in observational days (two-step GMM)

HAPPINESS	Model (a)	Model (b)	Model (c)	Model (d)	Model (e)
INCOME	0.001***	0.001***	0.002***	0.001***	0.001***
	(4.26)	(4.25)	(3.07)	(3.76)	(4.65)
UNEMPLOYMENT	−0.674***	−0.684***	−0.666**	−0.677***	−0.688***
	(−4.62)	(−4.85)	(−2.25)	(−4.51)	(−4.91)
HEALTH	−0.667***	−0.746***	−0.961***	−0.698***	−0.715***
	(−4.33)	(−4.97)	(−2.93)	(−4.40)	(−4.82)
SHOCK	−0.121***	−0.113***	−0.085	−0.111***	−0.111***
	(−4.32)	(−4.16)	(−1.53)	(−3.68)	(−4.12)
SEX	−0.672***	−0.647**	−0.547*	−0.594***	−0.663***
	(−4.50)	(−4.45)*	(−1.71)	(−3.86)	(−4.61)
TIMEDISCOUNT	0.004**	0.004**	0.007	0.003**	0.004**
	(2.37)	(2.57)	(1.49)	(2.15)	(2.50)
ANXIETY	0.114***	0.098***	0.060	0.105***	0.094**
	(3.00)	(2.65)	(0.66)	(2.72)	(2.56)
ALTRUISM	0.535**	0.437**	0.253	0.547**	0.480**
	(2.52)	(2.11)	(0.54)	(2.53)	(2.34)
SO_2 (MEAN)	38.552				
	(1.05)				
NO_2 (MEAN)		12.578			
	(0.80)				
CO (MEAN)			0.549		
		(0.60)			
Ox (MEAN)				−20.488	
			(−1.26)		
SPM (MEAN)					−8.470
				(−0.86)	
Constant	6.401***	6.275***	6.013***	6.762***	6.857***
	(16.81)	(10.20)	(5.87)	(17.20)	(15.47)
R^2	0.162	0.159	0.205	0.145	0.161
Observations	790	856	187	751	870

Note: Values in parentheses are t-values. ***, **, and * denote significance at 1%, 5%, and 10% levels, respectively.

Semi-parametric estimation

In this subsection, we apply a semi-parametric estimation. The model is as follows:

$$H_i = \rho_4 + \sigma_1 INCOME_i + \sigma_2 UNEMPLOYMENT_i + \sigma_3 HEALTH_i$$
$$+ \sigma_4 SHOCK_i + \sigma_5 SEX_i + \sigma_6 TIMEDISCOUNT_i + \sigma_7 ANXIETY_i$$
$$+ \sigma_8 ALTRUISM_i + \sigma_9 f(E_i^K) + \varepsilon_{4i}$$

Table 5.6 Estimation results: maximum concentration in observational days (two-step GMM)

HAPPINESS	Model (f)	Model (g)	Model (h)	Model (i)	Model (j)
INCOME	0.001***	0.001***	0.002***	0.001***	0.001***
	(4.29)	(4.25)	(3.09)	(3.72)	(4.65)
UNEMPLOYMENT	−0.673***	−0.686***	−0.675**	−0.670***	−0.686***
	(−4.61)	(−4.86)	(−2.28)	(−4.47)	(−4.90)
HEALTH	−0.672***	−0.737***	−0.954***	−0.701***	−0.720***
	(−4.36)	(−4.91)	(−2.92)	(−4.43)	(−4.85)
SHOCK	−0.121***	−0.113***	−0.085	−0.114***	−0.111***
	(−4.32)	(−4.17)	(−1.53)	(−3.78)	(−4.13)
SEX	−0.668***	−0.652***	−0.543*	−0.593***	−0.655***
	(−4.47)	(−4.48)	(−1.69)	(−3.86)	(−4.55)
TIMEDISCOUNT	0.004**	0.004***	0.007	0.003**	0.004**
	(2.38)	(2.60)	(1.49)	(2.09)	(2.46)
ANXIETY	0.113***	0.097***	0.062	0.105***	0.094**
	(2.98)	(2.62)	(0.68)	(2.72)	(2.57)
ALTRUISM	0.546**	0.436**	0.246	0.530**	0.482**
	(2.58)	(2.11)	(0.52)	(2.46)	(2.35)
SO_2 (MAX)	14.991				
	(0.85)				
NO_2 (MAX)		0.898			
		(0.08)			
CO (MAX)			0.233		
			(0.75)		
Ox (MAX)				−23.336**	
				(−2.04)	
SPM (MAX)					2.133
					(0.42)
Constant	6.416***	6.637***	6.045***	7.139***	6.514***
	(16.72)	(10.44)	(6.35)	(15.67)	(15.11)
R^2	0.161	0.159	0.206	0.148	0.161
Observations	790	856	187	751	870

Note: Values in parentheses are t-values. ***, **, and * denotes "significant" at 1%, 5%, and 10% level, respectively.

Here, ρ_4 is constant term and $f(\cdot)$ is a generic flexible functional form allowing for a potentially nonlinear non-monotonic relationship.

The estimation results are shown in Figures 5.8 through 5.17. An F-test shows that the estimation results of SO_2, NO_2, CO, and SPM are not statistically significant; we obtain statistically significant results only for Ox (mean and max). However, we are not able to interpret the case of the mean Ox concentration, because the confidence interval is too large to interpret. We therefore interpret only the result of the max Ox concentration (Figure 5.15). Figure 5.15 implies that degradation of the maximum Ox concentration results in a decrease in happiness. This result is consistent with that obtained in parametric estimations.

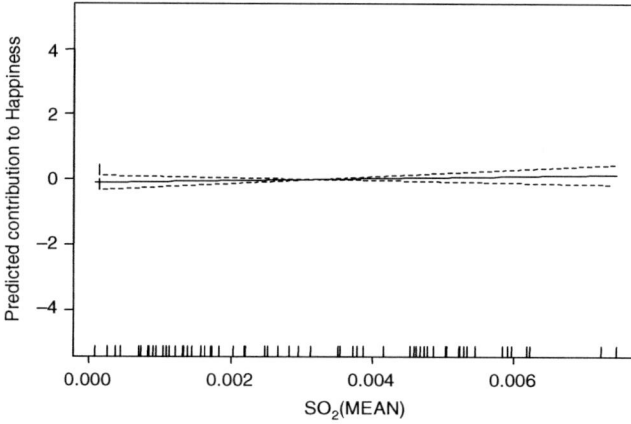

Figure 5.8 SO$_2$ (mean).

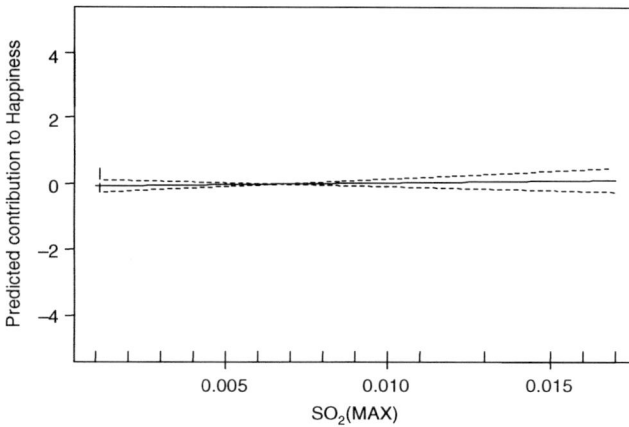

Figure 5.9 SO$_2$ (max).

To sum up, we obtain statistically significant estimation results only for the maximum Ox concentration both in parametric and semi-parametric estimations. Generally, the level of environmental degradation in Japan tends to be low, although photochemical oxidant smog does damage human health in some places. This may imply that only those environmental indices that have a bad effect on human health decrease the level of happiness, as shown in the subsection concerning environmental degradation and happiness.

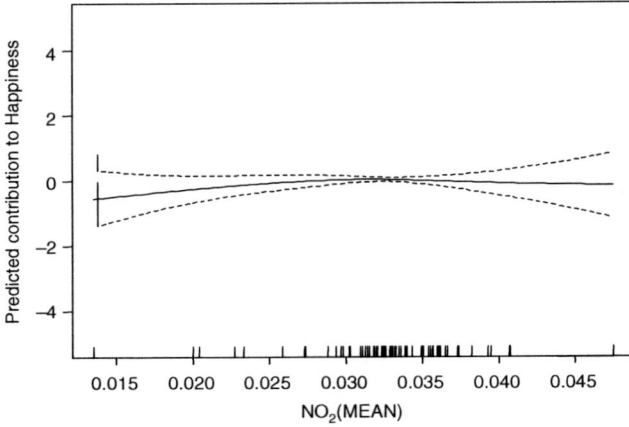

Figure 5.10 NO$_2$ (mean).

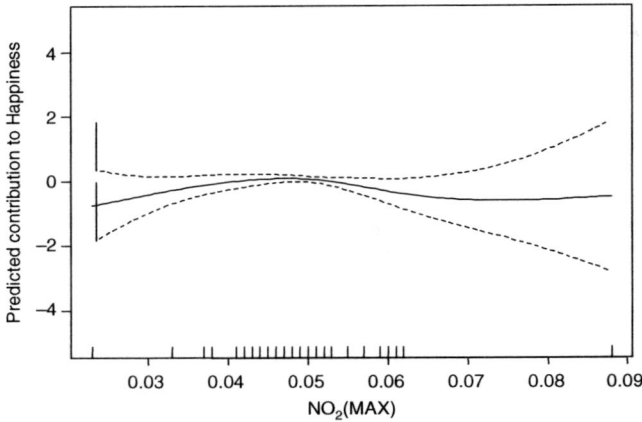

Figure 5.11 NO$_2$ (max).

Economic value of environmental improvement

In this subsection, we show the economic value of environmental improvement using the estimation results from the last subsection. Under a constant-utility criterion, by partially differentiating Equation (5.1) with respect to Ox, we obtain the following equation:

$$\frac{\partial INCOME_i}{\partial OX_i} = \frac{\partial_0}{\partial_1} = \frac{-23.336}{0.001} = 23336. \tag{5.1}$$

Equation (5.1) indicates a marginal trade-off between environmental degradation and income under a constant-utility condition, implying that the economic

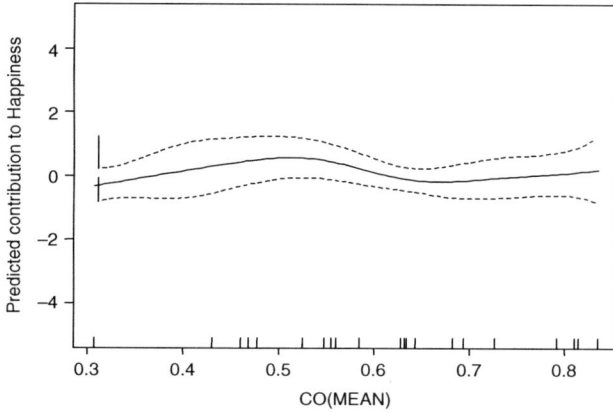

Figure 5.12 CO (mean).

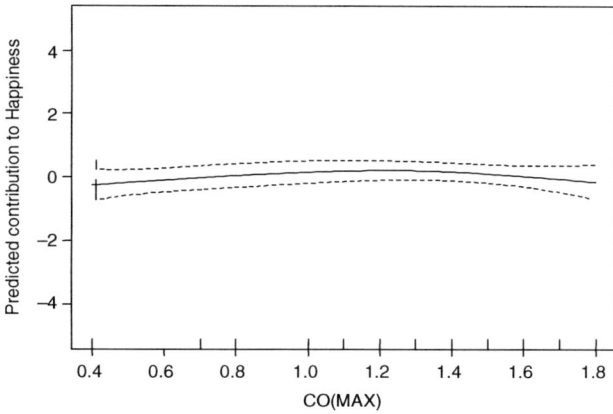

Figure 5.13 CO (max).

value of a 1-ppm change in Ox concentration is about 230 million yen. Because the sample average of the maximum Ox concentration is 0.0233 ppm, the economic value of a 1 percent decrease in Ox is 53,000 yen.

Happiness and attachment to nature: Country-level analysis

In the previous subsection, we considered the economic value of environmental improvement. The concept of a willingness to pay (WTP) for environmental conservation is used to evaluate environmental improvement. The person who has a strong attachment to nature seems to have a high WTP for environmental conservation. In this section, we explore whether a person who has a strong

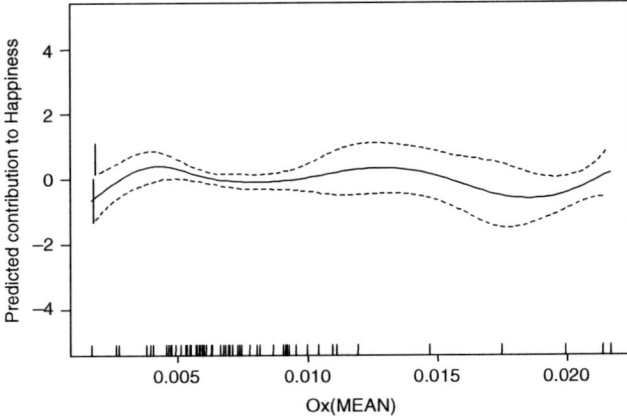

Figure 5.14 Ox (mean).

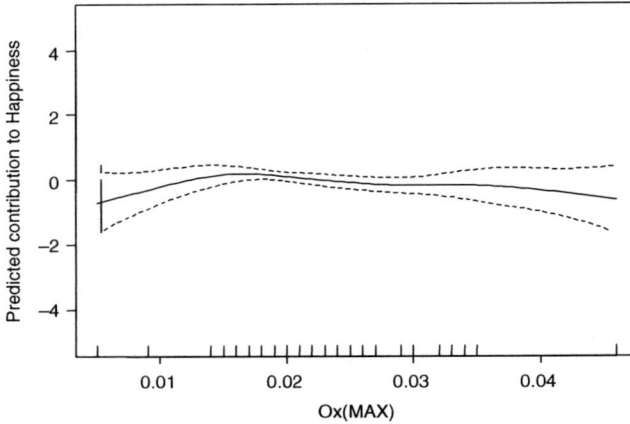

Figure 5.15 Ox (max).

attachment to nature – that is a person who has a high WTP for environmental conservation – is happier than others are. If so, environmental education should be adopted as a very valuable policy. Can having more contact with nature make people happier?

Survey outline

In this subsection, we explore the relationship between an individual's WTP (yen) for environmental conservation and happiness. To obtain an individual's WTP (yen), we conducted a survey in Japan. An Internet survey was conducted from 17 February to 20 February 2010. We give the survey outline and descriptive

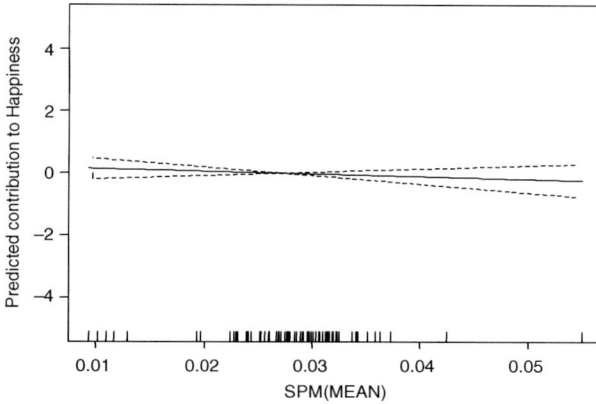

Figure 5.16 SPM (mean).

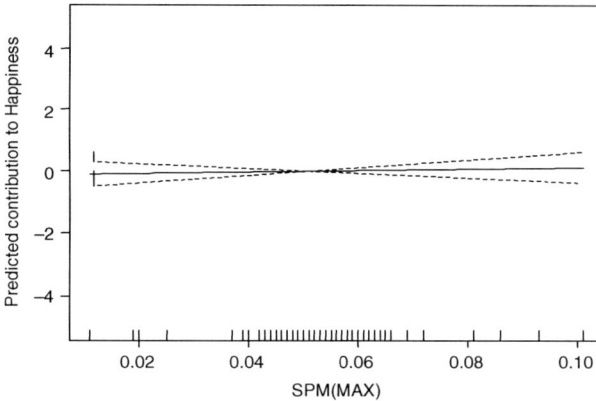

Figure 5.17 SPM (max).

statistics in Tables 5.7 and 5.8, respectively. Table 5.9 contains our questionnaire. Our survey questions are mainly based on Tsutsui *et al.* (2009).

To obtain an individual's WTP (yen) for environmental conservation, we apply a contingent valuation method (CVM), especially a single-bound dichotomous-choice method. We consider five scenarios: (a) biodiversity destruction due to a dam development project, (b) destruction of a water source forest, (c) water pollution (river), (d) agricultural damage due to global warming, and (e) wetland destruction. In this section, we consider the relationship between one's attachment to these environment situations and happiness.

To control for the effects of distance and time, we assume the place and time of destruction. Except for wetland destruction, we divide the survey sample into three groups in terms of place of destruction: (a) a local area, (b) a certain country in

Table 5.7 Survey outline

Period	17 February to 20 February 2010
Target	7,231 households in Japan
Method	Internet survey
Survey item	(1) Subjective well-being (HAPPINESS)
	(2) Household income (INCOME)
	(3) Disposable income (DI)
	(4) Unemployment unrest (UNEMPLOYMENT)
	(5) Age (AGE)
	(6) Sex (SEX)
	(7) Marriage (MARRIAGE)
	(8) Health unrest (HEALTH)
	(9) Number of shocking events (SHOCK)
	(10) Smoking (SMOKING)
	(11) Drinking (DRINKING)
	(12) Gambling (GAMBLING)
	(13) Competitive Spirit (COMPETITIVE)
	(14) Time discounting (TIMEDISCOUNT)
	(15) Risk Aversion (ANXIETY)
	(16) Altruism (ALTRIUM)
	(17) Willingness to pay (WTP)

Table 5.8 Descriptive statistics

	Average	*SD*	*Max*	*Min*	*Obs*
HAPPINESS	5.88895	2.090015	10	0	7231
INCOME	4.525239	1.954784	12	1	7231
DI	4.89061	2.47897	12	1	7231
UNEMPLOYMENT	0.2475453	0.431616	1	0	7231
AGE	44.57309	15.07368	86	13	7231
SEX	0.5452911	0.4979789	1	0	7231
MARRIAGE	0.6408519	0.4797839	1	0	7231
HEALTH	2.880514	1.088146	5	1	7231
SHOCK	2.217397	1.219837	5	1	7231
SMOKING	1.790071	1.491114	6	1	7231
DRINKING	2.777762	1.270033	6	1	7231
GAMBLING	2.345595	1.356576	6	1	7231
COMPETITIVE	3.261513	1.041496	5	1	7231
ANXIETY	6.592311	2.122596	11	1	7231
ALTRUISM	1.901812	0.767447	3	1	7231

Southeast Asia, and (c) a certain country in South America. In addition, we assume three time periods when the destruction occurs: (a) 5 years later, (b) 20 years later, and (c) 100 years later.

With regard to wetland destruction, we assume that the destruction occurs only in a local area. We further assume three kinds of scenarios in terms of the functions of the wetland to control for the functions of nature. The first scenario includes

Table 5.9 Questionnaire

	Question	Note
(1) INCOME	How much was your household income last year?	Yen
(2) UNEMPLOYMENT	Do you feel unrest about your own or your family's unemployment?	1: unrest 0: otherwise
(3) HEALTH	Do you feel unrest about your own or your family's health condition?	1: unrest 0: otherwise
(4) SHOCK	How many times have you had a shocking experience over the past 5 years?	Number of the experiences
(5) SEX	Gender	1: male 0: female
(6) TIMEDISCOUNT	How long can you wait for a person who is late for an appointment?	minutes
(7) ANXIETY	How much do you care about earthquake disaster?	0 (not care) to 10 (always care)
(8) ALTRUISM	Have you ever participated in volunteer activities?	Dummy (0: Yes, 1: No)
(9) HAPPINESS	All things considered, how satisfied or dissatisfied are you with your life as a whole now?	0 (dissatisfied) to 10 (satisfied)

the following condition: A wide variety of animals and plants live in the wetlands. The second scenario includes the following condition: A wide variety of animals and plants live in the wetland you evaluate. The wetland you evaluate also has several functions, such as cultural (for example, recreation), flood-control, water-retaining, and water-clarification functions. The third scenario includes the following condition: A wide variety of animals and plants live in the wetland you evaluate. The wetland you evaluate, however, doesn't have cultural, flood-control, water-retaining, and water-clarification functions. We summarize all scenarios in Table 5.10.

The model

The model is as follows:

$$H_i = \rho_5 + \pi_1 INCOME_i + \pi_2 DI + \sigma_3 UNEMPLOYMENT_i + \pi_4 AGE$$
$$+ \sigma_5 SEX_i + \pi_6 MARRIAGE_i + \pi_7 HEALTH_i + \pi_8 SHOCK_i$$
$$+ \pi_9 SMOKING_i + \pi_{10} DRINKING_i + \pi_{11} GAMBLING_i$$
$$+ \pi_{12} COMPETITIVE_i + \pi_{13} TIMEDISCOUNT_i + \pi_{14} ANXIETY_i$$
$$+ \pi_{15} ALTRUISM_i + \pi_{16} WTP_i^K + \varepsilon_{5i}.$$

Table 5.10 All scenarios

Environment	Functions	Where the destruction occurs	When the destruction occurs
Dam development	–	Local (Japan), Southeastern Asia, South America	5,20,100
Water source forest	–	Local (Japan), Southeastern Asia, South America	5,20,100
Water pollution (river)	–	Local (Japan), Southeastern Asia, South America	5,20,100
Agricultural damage by global warming	–	Local (Japan), Southeastern Asia, South America	5,20,100
Wetland 1	Wide variety of habitats	Local (Japan)	5,20,100
Wetland 2	Wide variety of habitats; many functions	Local (Japan)	5,20,100
Wetland 3	Wide variety of habitats; no functions	Local (Japan)	5,20,100

Here, DI_i denotes a household's disposable income; AGE_i denotes age; $MARRIAGE_i$ stands for the marriage dummy that takes a value of 1 if the person is married and 0 otherwise; $SMOKING_i$, $DRINKING_i$, and $GAMBLING_i$, represent their smoking, drinking, and gambling habits, respectively; $COMPETITIVE_i$ denotes the spirit of competitiveness; and WTP_i^k denotes a k-type WTP (yen) (k denotes scenarios). We separate our specification for each scenario to avoid multicollinearity. The other variables are already defined in the subsection on model and data.[7]

Estimation results

Table 5.11 shows the estimation results for a dam development project. We generally obtain the expected signs concerning economic and socio-demographic factors, psychological characteristics, and habitude indices. Concerning individuals' WTPs (yen), which are proxies for attachment to nature and our interests, we obtain statistically significant positive coefficients for all models. This implies that attachment to nature is positively related to happiness.

Table 5.12 summarizes the estimated coefficients for individuals' WTPs (yen) from all scenarios. We omit the estimated coefficients for economic and socio-demographic factors, psychological characteristics, and the habitude indices due

Table 5.11 Estimation results for the dam development project

	Model 1		Model 2		Model 3	
INCOME	0.047***	(3.20)	0.050***	(3.38)	0.051***	(3.49)
DI	0.127***	(11.98)	0.127***	(12.05)	0.128***	(12.14)
UNEMPLOYMENT	−0.581***	(−11.12)	−0.588***	(−11.24)	−0.584***	(−11.17)
AGE	−0.005***	(−2.98)	−0.005***	(−2.78)	−0.005***	(−2.89)
SEX	−0.705***	(−15.12)	−0.718***	(−15.40)	−0.721***	(−15.45)
MARRIAGE	0.707***	(13.11)	0.702***	(13.01)	0.704***	(13.04)
HEALTH	0.335***	(15.49)	0.334***	(15.44)	0.334***	(15.42)
SHOCK	−0.149***	(−8.06)	−0.148***	(−8.01)	−0.146***	(−7.89)
SMOKING	−0.106***	(−6.93)	−0.107***	(−6.99)	−0.109***	(−7.09)
DRINKING	0.022	(1.17)	0.021	(1.13)	0.021	(1.09)
GAMBLING	−0.078***	(−4.61)	−0.078***	(−4.63)	−0.078***	(−4.65)
COMPETITIVE	−0.145***	(−6.49)	−0.141***	(−6.31)	−0.141***	(−6.33)
TIMEDISCOUNT	−0.003***	(−3.00)	−0.003***	(−3.01)	−0.003***	(−3.10)
ANXIETY	−0.082***	(−7.81)	−0.083***	(−7.98)	−0.084***	(−8.06)
ALTRUISM	0.345***	(11.83)	0.357***	(12.31)	0.371***	(12.87)
dam_5	0.0262***	(3.12)				
dam_20			0.0321***	(3.01)		
dam_100					0.0316***	(2.36)
Constant	5.213***	(27.72)	5.250***	(27.92)	5.275***	(28.03)
R^2	0.225		0.224		0.222	
Observations	7231		7231		7231	

Note: Values in parentheses are z-values. ***, **, and * denote significance at 1%, 5%, and 10% levels, respectively. _5, _20, and _100 denote the time when the destruction occurs.

to space limitations. We obtain almost the same results for these coefficients. Table 5.12 implies that the coefficients for individuals' WTPs (yen) from local areas and Southeast Asia are statistically significant for almost all of the scenarios, while those of South America are not. This suggests that attachment to nature far from one's local area does not tend to be related to happiness. Then again, the coefficients for individuals' WTPs (yen) for future conservation tend to be statistically insignificant, suggesting that attachment to nature in the future does not tend to be related to happiness. Overall, our results suggest that nature closer to us in terms of time and distance is more related to happiness.

Table 5.13 shows the estimation results for wetlands. We show only the coefficients of WTPs for wetlands. We obtain statistically significant positive coefficients for all scenarios. This suggests that, apart from the functions of wetlands, attachment to wetlands is related to happiness.

Conclusions

In this chapter, we explored the determinants of happiness. We expected environmental degradation and attachment to nature to affect the level of happiness, and our results imply that environmental degradation above a certain threshold

Table 5.12 Summary of estimation results

	When the destruction occurs		
	5	20	100
Local area (Japan)			
Dum	0.0317** (2.39)	0.0438*** (2.72)	0.0341* (1.67)
Forest	0.0449*** (3.63)	0.0417*** (2.85)	0.0238 (1.29)
Water	0.0426*** (3.59)	0.0451*** (3.35)	0.0336* (1.96)
Agriculture	0.043*** (3.56)	0.0447*** (3.19)	0.0309* (1.73)
Southeastern Asia			
dum	0.0405*** (2.60)	0.0264 (1.30)	0.0277 (1.16)
Forest	0.0385*** (2.71)	0.0352* (1.95)	0.038* (1.70)
Water	0.0299** (2.21)	0.0332** (1.97)	0.0397** (1.97)
Agriculture	0.0324** (2.38)	0.0372** (2.16)	0.0418** (1.99)
South America			
Dum	0.00654 (0.43)	0.0187 (0.93)	0.031 (1.18)
Forest	0.0175 (1.22)	0.0134 (0.76)	0.0361 (1.60)
Water	0.011 (0.78)	0.0113 (0.65)	0.0211 (0.97)
Agriculture	−0.00268 (−0.19)	0.0132 (0.73)	0.0413* (1.74)

Note: Values in parentheses are z-values. ***, **, and * denote significance at the 1%, 5%, and 10% levels, respectively.

Table 5.13 Estimation results for wetland

	When the destruction occurs		
	5	20	100
Local area (Japan)			
Wetland 1	0.028*** (3.87)	0.0288*** (3.32)	0.0383*** (3.37)
Wetland 2	0.0298*** (4.27)	0.0377*** (4.60)	0.0471*** (4.47)
Wetland 3	0.0263*** (3.60)	0.0264*** (2.99)	0.0414*** (3.61)

Note: Values in parentheses are z-values. ***, **, and * denote significance at 1%, 5%, and 10% levels, respectively.

can affect people's level of happiness, and their attachment to nature tends to be positively related to happiness. We therefore conclude that both environmental conservation and attachment to nature can increase the level of people's happiness.

Our results suggest that if environmental education can improve people's attachment to nature, it would lead to increased happiness. In addition, although this is beyond the scope of this study, we can confidently say that if people who grow up close to nature tend to have a strong attachment for nature, living surrounded by nature will improve our happiness.

Notes

1 Easterlin (1974) presents an "Easterlin Paradox," which suggests that we cannot necessarily observe a correlation between income and happiness.
2 CO_2 emissions from fossil fuels are calculated following the method of Marland *et al.* (2000). $CO_{2i} = FC_i FO_i C_i$, where the three terms represent the net fuel production, the fraction of the fuel oxidized, and the carbon content of the fuel, respectively, and i represents solid, liquid, or gaseous fuels. With regard to cement manufacture, an equation analogous to the above equation is used to calculate CO_2 emissions from calcining $CaCO_3$ using the cement chemistry data.
3 This is equal to indigenous production plus imports, and stock changes minus exports, and fuels supplied to ships and aircraft engaged in international transport.
4 Although we tried to use more flexible nonparametric functions such as a generalized multiplicative model, the results are difficult to interpret.
5 When we used a LOESS function in place of a cubic spline function, the results were almost the same.
6 Smoothing parameters are determined to minimize generalized cross validation (GCV).
7 With regard to psychological (*COMPETITIVE, TIMEDISCOUNT, ANXIETY*, and *ALTRUISM*) and habitude (*SMOKING, DRINKING*, and *GAMBLING*) indices, we adopt the questionnaire of Tsutsui *et al.* (2009).

References

Cole, M. A. and Elliott, R. J. R., 2003, "Determining the trade-environment composition effect: The role of capital, labor and environmental regulations," *Journal of Environmental Economics and Management*, 46(3): 363–383.

Easterlin, R. A., 1974, "Does economic growth improve the human lot? Some empirical evidence," in P. A. David and M. W. Reder (eds), *Nations and Households in Economic Growth: Essays in Honor of Moses Abramowitz*, New York: Academic Press, pp. 89–125.

Frey, B. S. and Stutzer, A., 2001, *Happiness and Economics: How the Economy and Institutions Affect Well-Being*, Princeton: Princeton University Press.

Hastie, T. J. and Tibshirani, R. J., 1990, *Generalized Additive Models*, New York: Chapman and Hall.

Kahneman, D. and Deaton, A., 2010, "High income improves evaluation of life but not emotional well-being," *Proceedings of the National Academy of Sciences of the United States of America*, 107(38): 16489–16493.

Layard, R., 2011, *Happiness: Lessons from a New Science*, London: Penguin.

Marland, G., Boden, T. A. and Andres, R. J., 2000, "Global, regional, and national fossil fuel CO_2 emissions," in *Trends: A Compendium of Data on Global Change*, Carbon Dioxide Information Analysis Center, Oak Ridge National Laboratory, US Department of Energy, Oak Ridge, TN, USA.

Maslow, A., 1954, *Motivation and Personality*, New York: Harper and Row.

Nordhaus, W. and Tobin, J., 1972, *Is Growth Obsolete?* NBER General Series 96, New York: Columbia University Press.

Peiro, A., 2006, "Happiness, satisfaction and socio-economic condition: Some international evidence," *The Journal of Socio-Economics*, 35(2): 348–365.

Scitovsky, T., 1976, *The Joyless Economy? An Inquiry into Human Satisfaction and Dissatisfaction*, Oxford: Oxford University Press.

Stern, D. I., 2005, "Global sulfur emissions from 1850 to 2000," *Chemosphere*, 58, 163–175.

Stone, C. J., 1985, "Additive regression and other nonparametric models," *Annals of Statistics*, 13(2): 689–705.

Tella, R. D., MacCulloch, R. J. and Oswald, A. J., 2001, "Preferences over inflation and unemployment: Evidence from surveys of happiness," *The American Economic Review*, 91(1): 335–341.

Tsutsui, Y., Ohtake, F. and Ikeda, S., 2009, *The Reason Why You Are Unhappy*, ISER Discussion Paper, No. 630 (in Japanese).

Welsch, H., 2002, "Preferences over prosperity and pollution: Environmental valuation based on happiness surveys," *Kyklos*, 55(4): 473–494.

——, 2006, "Environment and happiness: Valuation of air pollution using life satisfaction data," *Ecological Economics*, 58(4): 801–813.

——, 2007, "Environmental welfare analysis: A life satisfaction approach," *Ecological Economics*, 62(3–4): 544–551.

6 Productivity analysis on ecosystems and biodiversity

Kei Kabaya and Shunsuke Managi

Introduction

The economic value of goods and services provided by ecosystems is progressively revealed by estimation using the contingent valuation method and conjoint analysis (TEEB 2010). However, this value is not yet reflected in GDP or national accounting, and there are few studies on the contribution of ecosystems and biodiversity to GDP. One example is the econometric analysis with the time series data of ecosystem services and GDP conducted by Richmond *et al.* (2007). In this analysis, the contribution to GDP of net primary production, which is perceived as an alternative to ecosystem services, was estimated using the Cobb-Douglas production function. As a result, it was demonstrated that a 1 percent increase in the rate of change of net primary production will raise the actual GDP growth rate by 0.09 percent.

Biodiversity, which supports ecosystem services, has positive impacts on net primary production by promoting the efficient use of resources, for example sunlight and nutrition. Similar to the above, however, there are few studies on the economic value of biodiversity itself and the contribution to GDP thereof. Di Falco and Chavas (2008) constructed an agricultural production model containing biodiversity as one of several factors, indicating that crop diversity had positive effects on short to mid-term production and it reduced the negative impacts of short rainfall on production. However, this is a case study limited to agricultural productivity within one region in one country. Considering that biodiversity is of global importance, which may affect not only agriculture but also genetic resources and ecotourism, further analysis from a wider perspective such as national GDP will be required.

In this chapter, the contribution of ecosystems and biodiversity to national GDP is analyzed. First of all, the framework of a productivity model reflecting the effects of ecosystems and biodiversity will be constructed. The data and econometric measures will be clarified as a next step, and then estimation results will be presented. Finally, the contribution of ecosystems and biodiversity to national GDP will be discussed through scenario analysis.

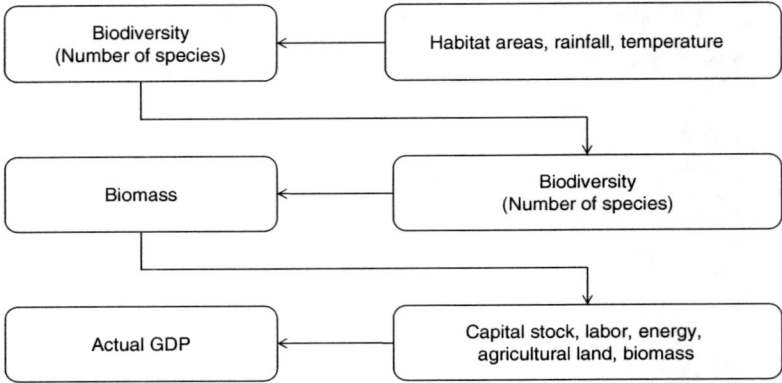

Figure 6.1 Overview of the econometric analytical model.

Econometric analytical model

The three-step estimation is conducted using this econometric model (see Figure 6.1). The effects of several factors – for example habitat areas – on biodiversity are scrutinized first and the result of this is utilized for the next estimation of biomass. Finally, the effect of biomass on GDP is analyzed with the Cobb-Douglas production function which reflects the preceding estimation results. In a series of estimations, the contribution to GDP of not only biomass but also habitat areas and biodiversity can be examined theoretically.

Biodiversity model

Biodiversity is here defined as species diversity. Data on the number of species in each country do exist, but reliable data obtained from a single source are only for one year. As this cannot be applied to productivity analysis which uses a time series dataset, the model to estimate the number species in each year is framed based on the assumption that this number changes year by year. Presumption of annual changes in the number of species seems unrealistic, but this is imaginable under the current situation where up to 40,000 species become extinct annually.

A habitat area affects the number of species in general. McGuinness (1984) proposes interesting hypotheses for the theoretical explanation thereof: the Habitat Diversity Hypothesis, the Equilibrium Theory (or Area *per se* Hypothesis), and the Disturbance Hypothesis. The first hypothesis suggests that larger areas will create different environmental habitats, which would generate a diversity of species. The second one is the application of island ecology, namely, species on a small island become extinct more quickly due to their small population. The last hypothesis is based on the ecological edge effect: the extinction rate might be accelerated in a small habitat because the ratio of an outer perimeter to an area is relatively large and disturbances from outside could occur more frequently.

The theoretical model on the relationship between habitat areas and diversity of species has already been constructed, and can be expressed as follows (Arrhenius 1921; Coleman *et al.* 1982):

$$S = wA^z \quad \therefore \ln S = \ln w + z \ln A$$

where S is the number of species, A is a habitat area, and w and z are constants. Since the number of species is correlated with rainfall and temperature (Gaston 2000), these variables are also integrated into the estimating equation as follows:

$$\ln S_{i,t} = c_1 + \alpha_1 \ln A_{i,t} + \alpha_2 \ln P_{i,t} + \alpha_3 \ln C_{i,t} + \varepsilon_{1,i,t}, \tag{6.1}$$

where P is rainfall, C is temperature, α is the coefficient of each variable, c is a constant, ε is an error term, and the subscripts i and t are country and year, respectively. The estimating equation itself assumes a time series, but t will take only one variable in this case, as only one year of data are available. The reason why a time series model is applied here is that time series data of the number of species can be generated as the estimation results, which could be utilized later in the following productivity model.

Biomass model

As different species use resources – for example light and water – in different ways, diversity of species promotes efficient use of resources, thereby improving productivity such as in net primary production (Tilman *et al.* 2005). Primary production is recognized as one of the supporting services derived from ecosystems (MA 2005a), which is a well-known example of the contribution of biodiversity to ecosystem services. Here, the relationship between biodiversity and primary production will also be estimated.

Biomass is used as a proxy for net primary production in the estimation, since the data of the latter are difficult to obtain. Biomass is defined as the volume of living organisms in the specific region, and is generally expressed with the weight of the relevant species. Although population density or environmental capacity should be taken into account in the analytical framework for the relationship between biodiversity and biomass (Cardinale *et al.* 2004), it is almost impossible to obtain such data on a national scale. Hence, the simple linear equation is assumed here as follows:

$$\ln B_{i,t} = c_2 + \beta \ln S_{i,t} + \varepsilon_{2,i,t}, \tag{6.2}$$

where B is biomass, β is the coefficient of each variable, and c, S, and ε have the same meaning as in Equation (6.1). As for S, the estimation results for Equation (6.1) will be applied, so that the indirect effects of habitat areas, rainfall, and temperature on biomass can also be analyzed. Similar to the biodiversity model, time series data could be generated from this estimation's results in spite of limited biomass data.

Productivity model

Biomass including foods and timbers can be regarded as natural resources in general, which can be incorporated in the production function as one factor with capital, labor, energy, and land. The estimating equation is framed here as follows:

$$\ln Y_{i,t} = c_3 + \gamma_1 \ln K_{i,t} + \gamma_2 \ln L_{i,t} + \gamma_3 \ln E_{i,t}$$
$$+ \gamma_4 \ln T_{i,t} + \gamma_5 \ln B_{i,t} + \varepsilon_{3,i,t},$$
$$\varepsilon_{3,i,t} = \eta_{3,i} + v_{3,i,t}, \tag{6.3}$$

where Y is the actual GDP, K is the capital stock, L is labor, E is energy input, T is agricultural land, γ is the coefficient of each variable, and the error term ε consists of country-specific effects η and disturbance v. The advantage of this panel model combining data across regions and over time is that controlling for individual effects enables extraction of common effects on the dependent variable, and increasing the number of observations improves multicollinearity and degrees of freedom, thereby enhancing unbiasedness (Kitamura 2005).

Panel estimation can be the basis of an analytical model for temporal dynamism by adding a lagged dependent variable. The merits of dynamic panel analysis are that inter-temporal changes of the same economic actor can be empirically examined, and that future variation and policy response can be predicted from the current trends (Kitamura 2005). From this aspect, Equation (6.3) can be converted as follows:

$$\ln Y_{i,t} = c_4 + \delta_0 \ln Y_{i,t-1} + \delta_1 \ln K_{i,t-1} + \delta_2 \ln L_{i,t} + \delta_3 \ln E_{i,t}$$
$$+ \delta_4 \ln T_{i,t-1} + \delta_5 \ln B_{i,t} + \eta_{4,i} + v_{4,i,t}, \tag{6.4}$$

where δ denotes the coefficient of each variable. Actual GDP at time $(t-1)$ is supplemented as an explanatory variable in this model so as to estimate the effects of actual GDP in the previous year. Also, $(t-1)$ is selected for the capital stock K and agricultural land T, with a view to reflecting the nature of economic activities: the stocks of capital and developed land by time $(t-1)$ would be inputs into production at time t.

Furthermore, Equation (6.4) will be transformed to the first-difference model, as follows, in order to eliminate the individual effects and improve estimation accuracy by reducing serial correlation (Baltagi 2001):

$$\Delta \ln Y_{i,t} = \sigma_0 \Delta \ln Y_{i,t-1} + \sigma_1 \Delta \ln K_{i,t-1} + \sigma_2 \Delta \ln L_{i,t} + \sigma_3 \Delta \ln E_{i,t}$$
$$+ \sigma_4 \Delta \ln T_{i,t-1} + \sigma_5 \Delta \ln B_{i,t} + v_{5,i,t}, \tag{6.5}$$

where σ is the coefficient of each variable, and Δ indicates a first-difference. The variable of $\Delta \ln B$ denotes the increase in biomass from year $(t - 1)$ to year t, which may imply net primary production.

Data and estimation method

Data

As for time series data for economic indicators, habitat areas, and climate indexes in each country, a certain timeframe is set to construct panel data. Here, datasets in 66 countries from 1990 to 2000 will be targeted in the light of data availability (see Table 6.1). The datasets of actual GDP, the capital stock, and labor are obtained from the Extended Penn World, Table 3.0, which is reliable at the international level. The World Development Indicators organized by the World Bank provide the data of energy input, expressed as oil-equivalent conversion, and the FAOSTAT offers the data on agricultural land as well as habitat areas, specifically forests and inland water. The number of species is the aggregated data of the number of animal and plant species obtained from the Biodiversity Data Sourcebook provided by the World Conservation Monitoring Center, and the data for biomass (defined as the tree volume upper ground) are acquired from the FAO (2001). As for rainfall and temperature, the data are collected from the Tyndall Centre for Climate Change Research, which keeps national yearly datasets from 1900 to 2000. This information is organized in Table 6.2 with the basic statistics.

Estimation method

Pooled Ordinary Least Squares (OLS) will be applied to Equation (6.1), as it is linear and consists of cross-section datasets. More precisely, the forward-selection method, one of the step-wise methods, will be used here to extract only the significant variables and to calculate the estimated results which could be utilized in the next estimation. Equation (6.2) will be analyzed with Two-Stage Least Squares (2SLS) to reflect the previous results. Excluded variables in the first estimation are also integrated into this estimation because they may have significant impacts on biomass.

As for Equation (6.5), which incorporates the second set of estimation results, fixed effects estimation considering variance in the constant due to national uniqueness, as well as random effects estimation assuming a randomly determined constant, will be applied in addition to the OLS. Moreover, as endogenous variables may be included in Equation (6.5), it will be also examined with the Generalized Method of Moments (GMM). The system GMM, which reduces bias from fixed effects by transforming a first-difference model, and which corrects endogeneity problems by utilizing lagged endogenous variables as instruments, will be applied (Blundell and Bond 1998). At the time of estimation, all explanatory variables, including lagged dependent variables, will be perceived as endogenous variables. While both one-year and two-year lags will be utilized for estimation in differences, only the two-year lag will be used in the levels. Furthermore, year,

Table 6.1 Countries included in the analysis

Country		
Asia region	Honduras	**Middle East**
Bangladesh	Mexico	Iran
Brunei Darussalam	Nicaragua	Iraq
China	Panama	Jordan
D.P.R.Korea	Peru	Syria
India	**Europe**	**Africa**
Indonesia	Denmark	Benin
Japan	Finland	Botswana
Malaysia	France	Cameroon
Nepal	Germany	Congo
Philippines	Greece	D. R. Congo
Republic of Korea	Hungary	Egypt
Sri Lanka	Iceland	Kenya
Thailand	Ireland	Morocco
Latin America	Italy	Mozambique
Argentina	Netherlands	Nigeria
Bolivia	Norway	Senegal
Brazil	Poland	South Africa
Chile	Portugal	Sudan
Colombia	Spain	Togo
Cuba	Sweden	Tunisia
Ecuador	Switzerland	Tanzania
El Salvador	United Kingdom	Zambia
Guatemala		Zimbabwe

regional dummy, national land area, and an OECD dummy are added to the instruments as exogenous variables. Finally, two-step estimation will be conducted with the heteroskedasticity-robust standard error.

Results of the econometric analysis

Biodiversity model

The estimation result for Equation (6.1) with the forward-selection method is indicated in Table 6.3. As the F-statistic was significant at the 0.01 level, and the adjusted R^2 was large enough, this biodiversity model can be concluded to be significant and robust. Also, the estimation result demonstrated the significance of three variables, namely, forest areas, rainfall, and temperature at the 0.01 level. Positive impacts of forest areas on species diversity are consistent with the habitat area-species theory: an increase in forest areas of 1 unit will increment biodiversity by 0.28 units here. Likewise, the positive impacts of rainfall and temperature on biodiversity are compatible with the theory on the relationship between available resources and species diversity (Wright 1983; Turner *et al.* 1988). An increase in each unit of annual average precipitation as well as annual average temperature

Table 6.2 Basic statistics

Variable	Definition	Unit	Observation	Average	Standard deviation	Minimum	Maximum	Source
Y	Actual GDP (PPP in 2000)	International dollar (10^6)	726	349,322	662,341	3,219	5,052,200	a
K	Capital stock (PPP in 2000)	Intl. dollar (10^6)	726	528,412	1,165,227	2,662	7,813,245	a
L	Employed labor	1,000 persons	726	31,044	99,003	110	755,338	a
E	Energy consumption (oil equivalent)	1,000t	726	70,862	148,925	720	1,092,154	b
T	Land area (agricultural land)	1,000ha	726	34,727	76,525	10	544,862	e
S	Biodiversity (fauna + flora)	Species	66	9,226	11,435	404	59,214	c
B	Biomass (wood volume)	(10^6)	66	3,999	14,513	1	113,676	d
F	Forest area	1,000ha	726	29,995	73,929	9	574,839	e
W	Inland water area	1,000ha	726	2,366	5,463	13	31,407	e
P	Average annual precipitation	mm	726	1,172	746	37	3,606	f
C	Average annual temperature	Degrees Celsius	726	18.4	7.8	1.3	28.9	f

Sources:

(a) Extended Penn World Tables 3.0: http://homepage.newschool.edu/~foleyd/epwt/
(b) World Development Indicators: http://data.worldbank.org/indicator/EG.USE.COMM.KT.OE
(c) Biodiversity Data Sourcebook: http://www.unep-wcmc.org/resources/publications/biodiv_series.htm
(d) FAO Global Forest Resources Assessment 2000: http://www.fao.org/forestry/fra/2000/report/en/
(e) FAOSTAT: http://faostat.fao.org/site/377/default.aspx#ancor
(f) TYN CY 1.1 Tyndall Centre for Climate Change Research: http://www.cru.uea.ac.uk/~timm/cty/obs/TYN_CY_1_1.html

Note: International dollars are defined asthe currency used when reporting purchasing poer parity (PPP) data.

Table 6.3 Estimation result of biodiversity model

Variable	Biodiversity ($\ln S$)	
	Coefficient	t-value
$\ln F$	0.27729	6.28***
$\ln W$	–	–
$\ln P$	0.37901	3.62***
$\ln C$	0.37945	3.03***
Constant	2.52301	3.39***
Observations	66	
F-statistic	31.25***	
Adjusted R^2	0.5827	
VIF	1.09	

***significant at the 0.01 level; **significant at the 0.05 level; *significant at the 0.10 level.

Table 6.4 Estimation result of biomass model

Variable	Biomass ($\ln B$)	
	Coefficient	t-value
$\ln S$	2.24091	14.32***
$\ln W$	0.32136	4.51***
Constant	−15.08967	−12.07***
Observations	66	
F-statistic	162.21***	
Adjusted R^2	0.8322	

***significant at the 0.01 level; **significant at the 0.05 level; *significant at the 0.10 level.

will contribute to the growth of biodiversity by 0.38 units each. It may seem unexpected that inland water areas have insignificant effects on biodiversity, but this can be regarded as reasonable when species diversity is grasped at the national scale, as inland water ecosystems hold only 2.4 percent of the recorded species in the world (MA 2005b).

Biomass model

The next estimation results from the 2SLS, reflecting the estimated outcome of Equation (6.2), are shown in Table 6.4. Similar to the biodiversity model, the F-statistic was significant and the adjusted R^2 was high enough, and so this model can be said to estimate biomass significantly as well as robustly. Glancing at each variable, both biodiversity and inland water areas affected biomass significantly at the 0.01 level. The positive impact of biodiversity on biomass is consistent with the theory suggested above: an increment of biodiversity of 1 point will increase biomass by 2.24 points. This result indicates that forest areas, rainfall, and temperature will also affect biomass through biodiversity indirectly: a rise in each 1 unit will increase biomass by 0.62, 0.85, and 0.85, respectively. The elasticity of inland water areas was 0.32, which was positive but comparatively smaller than for the other variables.

Productivity model

The estimation results from Equation (6.5) with the OLS, fixed effects estimation, and random effects estimation are set out in Table 6.5. As the F-statistic of the fixed effects estimation was significant and the null hypothesis was rejected by the Hausman test, the result of the fixed effects estimation is the next focus (Tsutsui *et al.* 2007). According to this result, significant independent variables include actual GDP and the capital stock at time $(t - 1)$, as well as the energy input and biomass at time t. The significant effect of actual GDP at time $(t - 1)$ on the one at time t implies that this model reflects economic dynamism well. However, it may be against our understanding that labor did not have a significant impact on actual GDP, and it seems that every estimated parameter was smaller than our expectation of it. This may be because biases were generated on the estimation model, for example, by lack of consideration of the endogeneity of lagged actual GDP.

Table 6.5 Estimation result of productivity model (1)

Variable	GDP growth rate ($\Delta \ln Y_t$)		
	OLS	Fixed effects	Random effects
$\Delta \ln Y_{t-1}$	1.00434	0.88339	1.00372
	(467.97***)	(43.66***)	(489.33***)
$\Delta \ln K_{t-1}$	0.15071	0.17996	0.14366
	(1.26)	(2.26**)	(2.50**)
$\Delta \ln L_t$	0.44790	−0.56373	0.35533
	(1.49)	(−1.18)	(1.34)
$\Delta \ln E_t$	0.31737	0.23906	0.30857
	(4.74***)	(4.17***)	(5.61***)
$\Delta \ln T_{t-1}$	0.12341	0.14906	0.12842
	(1.67*)	(1.31)	(1.16)
$\Delta \ln B_t$	−0.00378	0.06556	−0.00347
	(−2.53**)	(3.69***)	(−2.29**)
Constant	0.01557	0.95763	−0.00814
	(−0.68)	(3.79***)	(−0.32)
Observations	594	594	594
Number of countries	–	66	66
F-value/*Wald chi²*-value	.***	353.21***	334615.99***
R^2	0.9988	0.9910	0.9988
F-statistic	–	2.22***	–
Breusch and Pagan test	–	–	1.98
Hausman test	–	58.43***	

(·) denotes t-value/z-value; ***significant at the 0.01 level; **significant at the 0.05 level *significant at the 0.10 level.

Additionally, the estimation result of Equation (6.5) with the system GMM is shown in Table 6.6. The null hypothesis of no serial correlation could not be rejected by the Arellano-Bond test (AR(2)), which indicates that the absence of autocorrelation on the disturbance $v_{5,i,t}$ was supported by the data. Likewise, the null hypothesis of no correlation between instrumental variables and residuals could not be rejected by the Hansen J test, which satisfied the orthogonality conditions required by the system GMM. Hence, the estimation result of the system GMM is prioritized here. According to this estimation result, the explanatory variables significantly affecting the independent variable are actual GDP and the capital stock at time $(t-1)$ as well as labor, the energy input, and biomass at time t. As the logged difference $(\Delta \ln)$ can be perceived as an approximation of an elongation percentage period-over-period (Yamamoto 1992), a more precise interpretation of the result was that the elongation percentage period-over-period of the explanatory variables has significant impacts on the GDP growth rate. Each coefficient implied that the GDP growth rate at time $(t-1)$ would decelerate the GDP growth rate at time t, which may have reflected growth dynamism. Conversely but consistent with economic theories, the elongation percentage of the capital stock, labor, and the energy input will have positive impacts on the actual GDP growth rate. Likewise, an increased rate of biomass growth will contribute to it: the increment of the former by 1 percent will raise the latter by 0.045 percent. This relationship may originate from an increase in natural resources which could be an input into economic activities. However, its elasticity seemed comparatively smaller than of other variables.

Agricultural land was not significant in either the fixed effects estimation or the system GMM; therefore, further analysis excluding this variable was performed using the system GMM. As a result of a similar estimation to the above, which

Table 6.6 Estimation result of productivity model (2)

Variable	GDP growth rate ($\Delta \ln Y_t$)	
	Coefficient	t-value
$\Delta \ln Y_{t-1}$	−0.11254	−1.81*
$\Delta \ln K_{t-1}$	0.18114	2.06**
$\Delta \ln L_t$	0.81711	3.26***
$\Delta \ln E_t$	0.36174	5.53***
$\Delta \ln T_{t-1}$	0.03048	0.30
$\Delta \ln B_t$	0.04459	1.99**
Observations	594	
Number of countries	66	
F-statistic	171.30***	
AR(1)	−1.64 ($p=0.101$)	
AR(2)	−0.05 ($p=0.958$)	
Hansen J test	63.29 ($p=1.000$)	

***significant at the 0.01 level; **significant at the 0.05 level; *significant at the 0.10 level.

Table 6.7 Estimation result of productivity model (3)

Variable	GDP growth rate ($\Delta \ln Y_t$)	
	Coefficient	t-value
$\Delta \ln Y_{t-1}$	−0.11373	−1.80*
$\Delta \ln K_{t-1}$	0.17984	2.01**
$\Delta \ln L_t$	0.82304	3.53***
$\Delta \ln E_t$	0.35305	5.12***
$\Delta \ln B_t$	0.04101	1.93*
Observations	594	
Number of countries	66	
F-statistic	123.47***	
AR(1)	−1.64 ($p = 0.102$)	
AR(2)	−0.05 ($p = 0.957$)	
Hansen J test	57.24 ($p = 1.000$)	

***significant at the 0.01 level; **significant at the 0.05 level; *significant at the 0.10 level.

further added agricultural land to the instruments, every variable was significant at the level of 0.01, 0.05, or 0.10, thus satisfying the conditions required for the system GMM (see Table 6.7). Although the *F*-statistic is smaller than that of the previous estimation, this model is robust enough and is simpler than the previous one. Hence, this model will be applied for the subsequent scenario analysis.

Scenario analysis

Scenario setting

Based on the above three estimation results, a scenario analysis will be conducted here. The purpose of this is to show the impacts of ecosystems and biodiversity on national GDP with concrete figures, thereby considering their values. Taking advantage of this multistage estimation, two scenarios regarding independent variables for the biodiversity model – namely, forest areas, and rainfall and temperature – will be set, and their impacts on biodiversity, biomass, and actual GDP will be investigated. These variables are relevant to the issues of biodiversity and climate change, so the impact analysis may be meaningful for forest conservation and for climate-change mitigation.

Prior to the above scenario setting, the business-as-usual (BAU) scenario will be framed as follows:

- The scenario timeframe is set ten years from 2001.
- The annual rates of change of forest areas, capital stock, labor, and the energy input are randomly selected based on each normal distribution model with each average and standard deviation of the historical records from 1990 to 2000.

- Likewise, the figures for rainfall and temperature are randomly selected, based on each normal distribution.
- As for inland water areas, few significant changes were observed from 1990 to 2000; thus, no changes within this scenario timeframe are assumed.

Then, two scenarios will be set as follows:

- Scenario 1 (forest conservation): the average annual rates of change of forest areas will be decreased linearly and halved at the time of the tenth year, compared to the historical records from 1990 to 2000 in the case where they show negative trends during this period.
- Scenario 2 (unstable climate): the variance of historical records of rainfall and temperature will be doubled in the random selection of figures.
- Scenario 1 will investigate the contribution of forest areas, as well as biodiversity, to actual GDP; and Scenario 2 will analyze the impact of unstable climate on the GDP growth rate. Note that these two scenarios are not different from the BAU scenario except the above specific points.
- The estimates for each dependent variable calculated from each scenario will be incorporated into the biodiversity, biomass, and productivity models, which apply the above estimation results for each coefficient as follows:

$$\ln S_{i,t} = 2.52301 + 0.27729 \ln F_{i,t} + 0.37901 \ln P_{i,t} + 0.37945 \ln C_{i,t} \tag{6.1'}$$

$$\ln B_{i,t} = -15.08967 + 2.24091 \ln S_{i,t} + 0.32136 \ln W_{i,t} \tag{6.2'}$$

$$\ln Y_{i,t} = -0.11373 \Delta \ln Y_{i,t-1} + 0.17984 \Delta \ln K_{i,t-1} + 0.82304 \Delta \ln L_{i,t}$$
$$+ 0.35305 \Delta \ln E_{i,t} + 0.04101 \Delta \ln E_{i,t} \tag{6.5'}$$

Scenario analysis result

The overall trends of forest areas, number of species, biomass, and actual GDP during the above period in both the BAU scenario and Scenario 1 are shown in Figure 6.2. Every case demonstrates slightly higher scores for Scenario 1, which are indicated by the right bar of "difference" of Scenario 1 from the BAU scenario on each figure.[1] In terms of actual GDP, the differences between the two scenarios in each region are also shown in Figure 6.3: the highest score was on Latin America, followed by Africa and Asia, while Europe and the Middle East indicated no difference between the two scenarios. This may imply that forest conservation is urgent in the former three regions from economic aspects, while it is not such a big issue in the latter two regions as deforestation has not yet been observed.

The values of unit forest area and species were further investigated on the assumption of constant returns to scale. The former was calculated by the equation of the difference of actual GDP between two scenarios divided by the difference in forest areas, showing that the value tended to increment as time elapsed. This may be because forest areas have positive impacts, not on actual GDP, but on the GDP growth rate through biodiversity and biomass. From regional aspects, Asia showed a much higher value of unit forest area; approximately 319 international dollars[2] in the tenth year, followed by Latin America; approximately 112 international dollars; and Africa with approximately 66 international dollars (see Figure 6.4). A similar calculation was performed on the value of one species, which resulted in the highest value in Asia: approximately 2,984 international dollars at the tenth year; followed by Latin America: approximately 1,510 international dollars; and Africa: approximately 590 international dollars (see Figure 6.5). At the national level, Bangladesh held the highest forest value: approximately

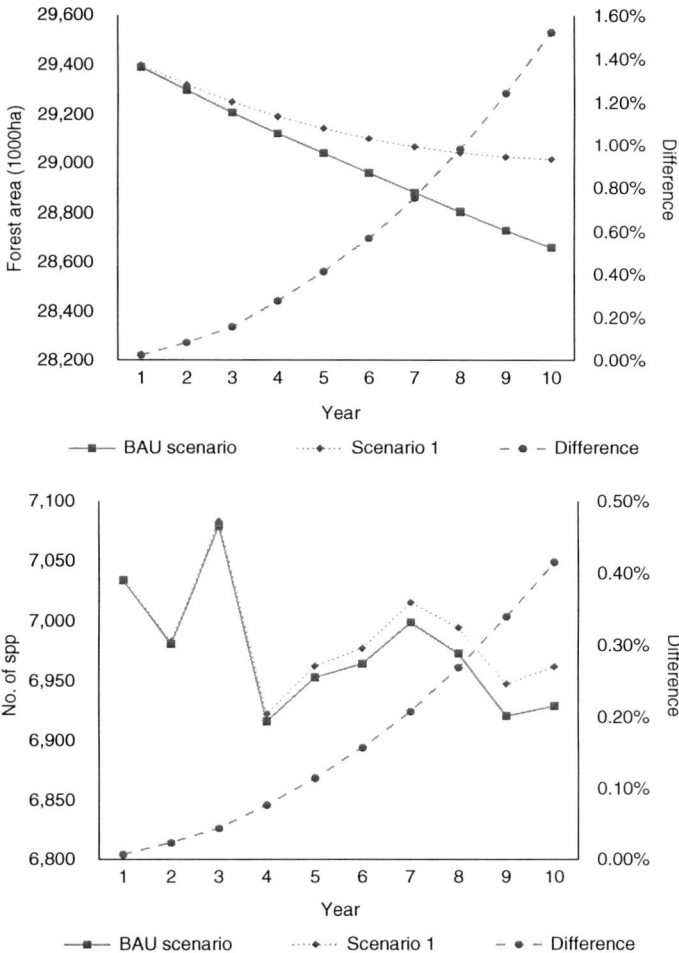

Figure 6.2 Difference between BAU scenario and Scenario 1 (overall trends).

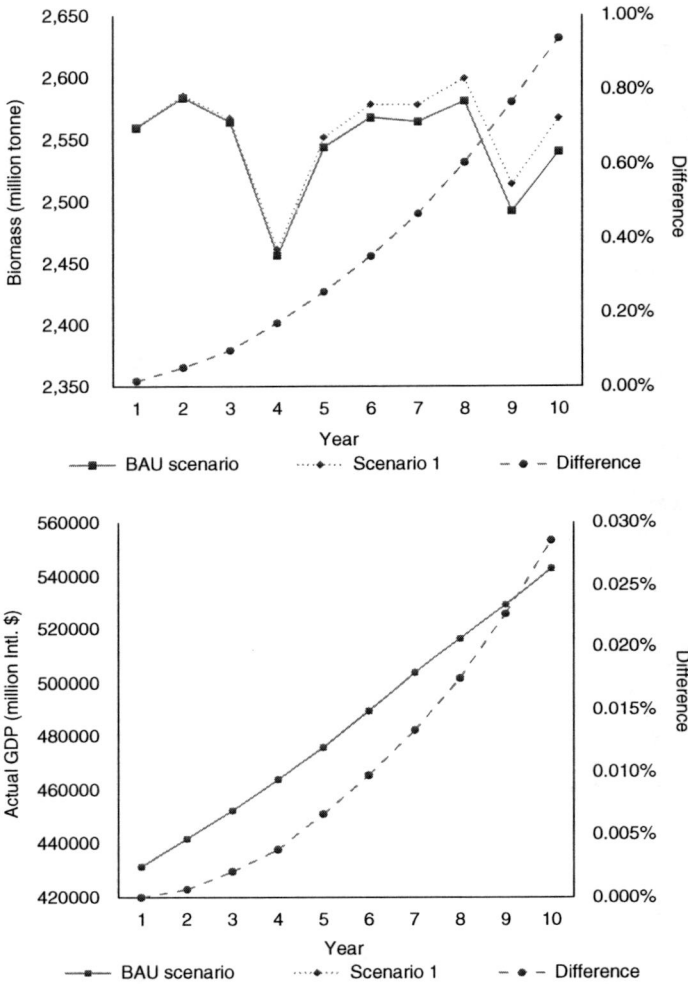

Figure 6.2 Continued.

4,675 international dollars at the tenth year; Japan marked the highest species value with approximately 27,903 international dollars.

The analytical results for Scenario 2 in terms of overall trends are demonstrated in Figure 6.6. As the variances of the independent variables of rainfall and temperature were doubled, fluctuations thereof became larger than in the BAU scenario. Subsequent to these trends, the number of species, biomass, and the GDP growth rate showed similar larger fluctuations. With regard to the GDP growth rate in each region, the Middle East showed large swings over time from 0.21 percent to

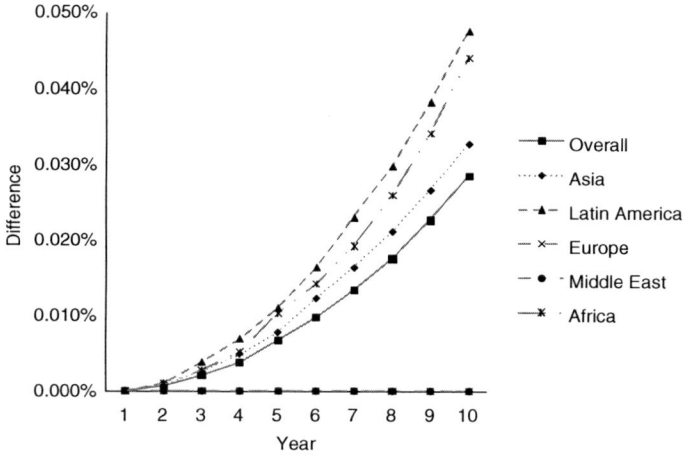

Figure 6.3 Difference in actual GDP between BAU scenario and Scenario 1 (each region).

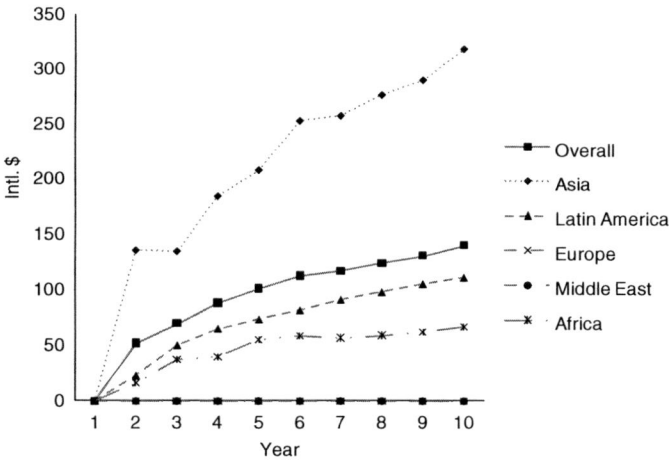

Figure 6.4 Economic contribution of unit forest area to actual GDP (each region).

9.45 percent (see Figure 6.7), which may imply the vulnerability of this region to climate change. On the other hand, Asia and Latin American were not so much affected by this scenario: from 1.61 percent to 4.22 percent, and from 2.67 percent to 4.46 percent, respectively.

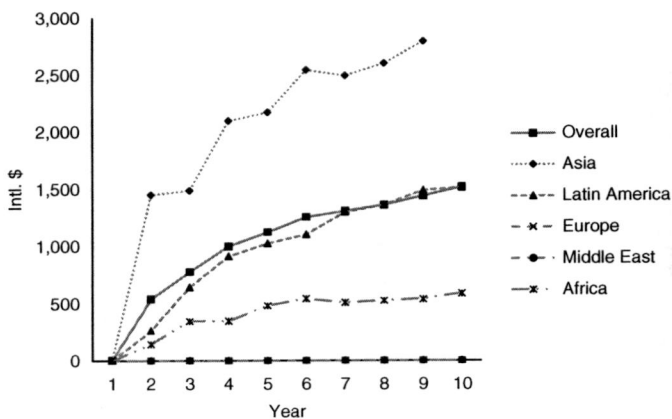

Figure 6.5 Economic contribution of species to actual GDP (each region).

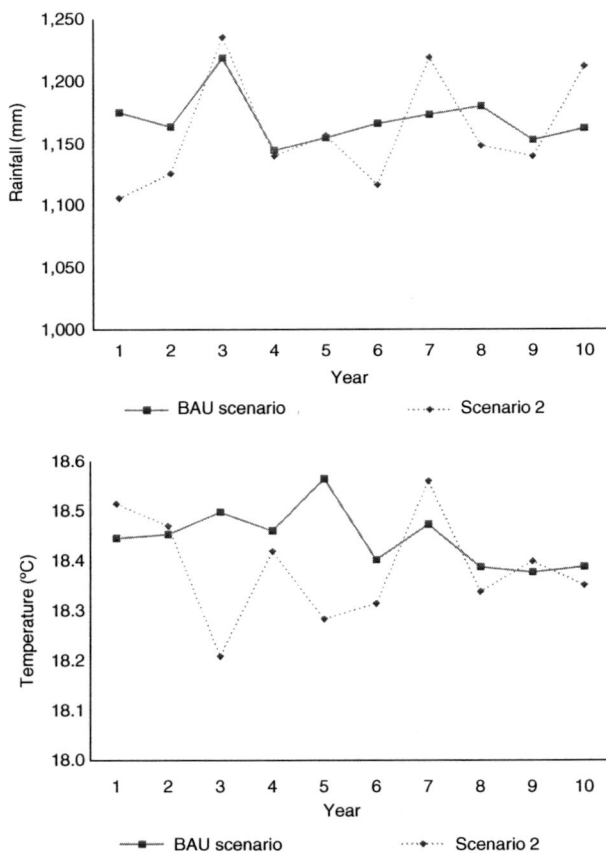

Figure 6.6 Difference between BAU scenario and Scenario 2 (overall trends).

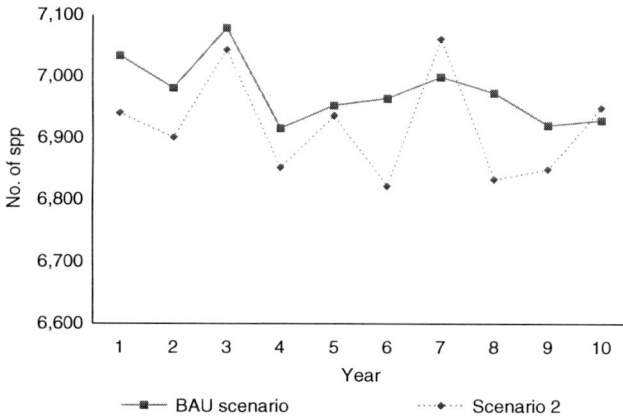

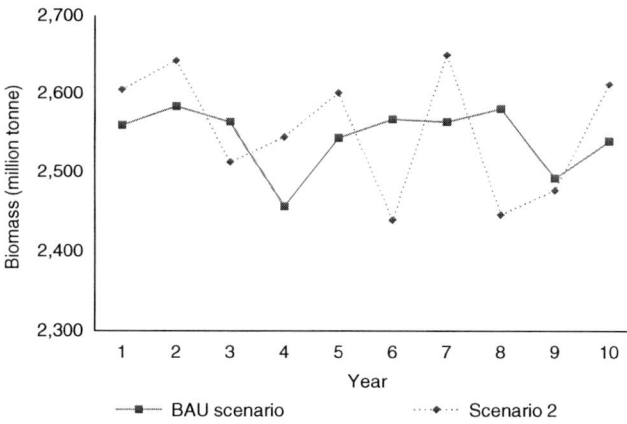

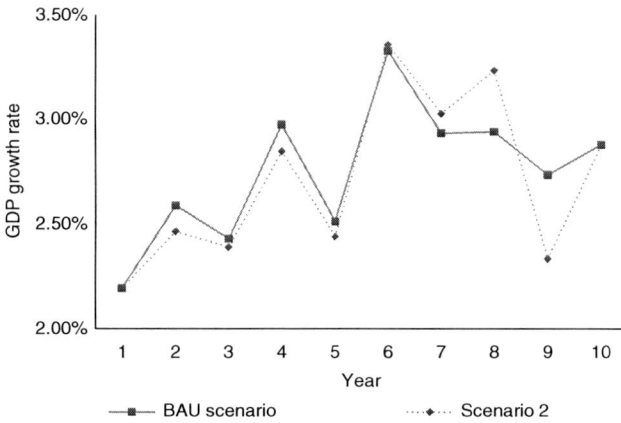

Figure 6.6 Continued.

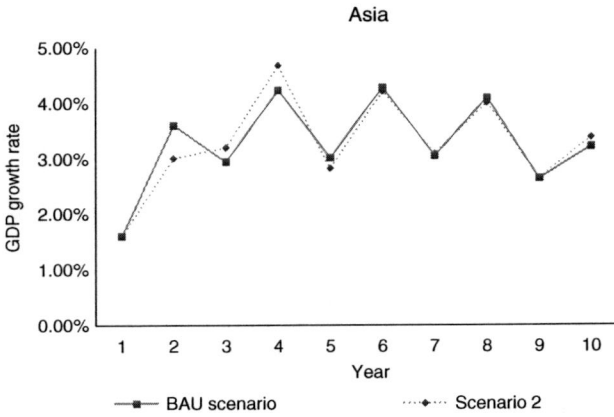

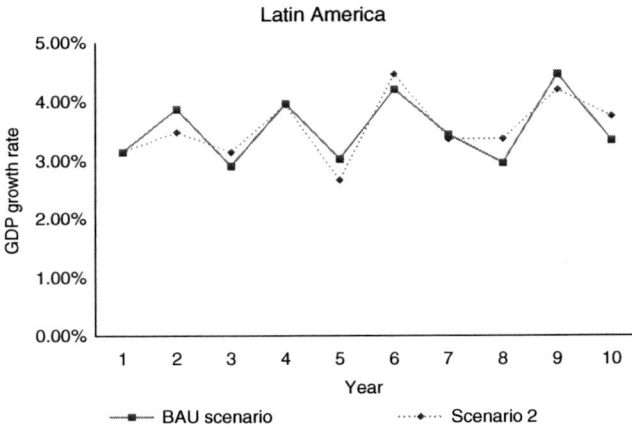

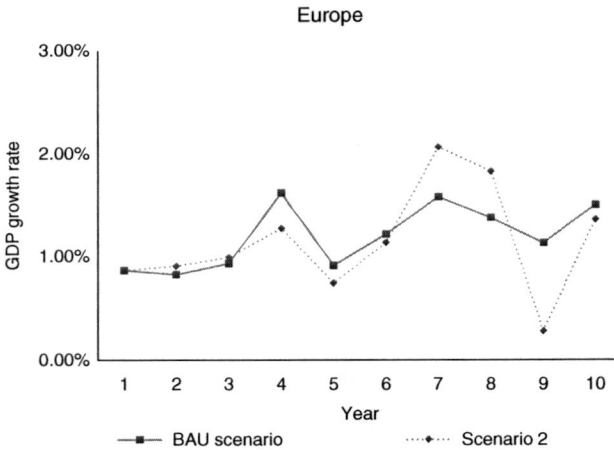

Figure 6.7 GDP growth rate in each scenario (each region).

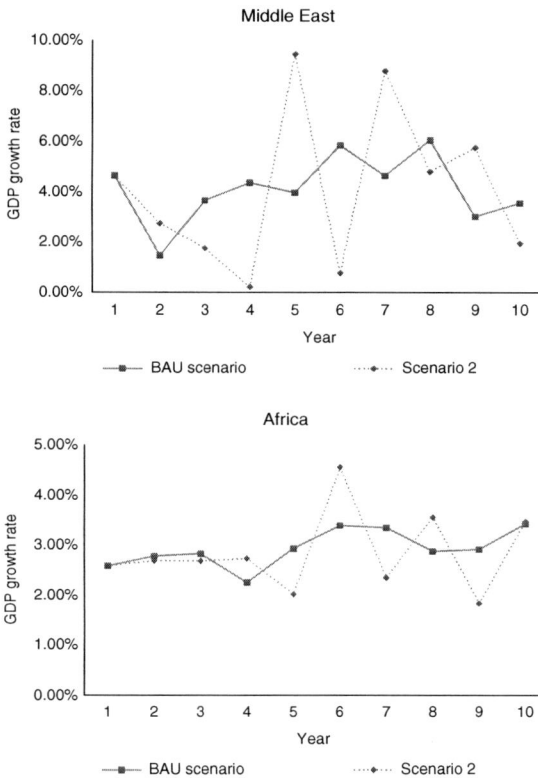

Figure 6.7 Continued.

Conclusion

Econometric analysis on biodiversity, biomass, and productivity models revealed that biodiversity which was estimated from forest areas, rainfall, and temperature would increase biomass in conjunction with inland water areas, which in turn would have positive, significant impacts on the GDP growth rate. Also, scenario analysis demonstrated that unit forest area as well as species could have the highest values in Asia, while unstable climate conditions defined as larger fluctuations of rainfall and temperature could have the biggest impacts in the Middle East.

In this analysis, the productivity of ecosystems and biodiversity was investigated through net primary production, and their values were estimated from the aspect of their contribution to actual GDP. Economic valuation from production, or the supply side, may complement that from consumption, the demand side, and may contribute to a detailed and accurate economic valuation. In this sense, economic values based on both supply and demand will need to be applied when those are utilized in future green accounting schemes.

Notes

1 Species do not fluctuate in such a short term in reality; therefore, this seems a limitation of this econometric model.
2 International dollars are the currency used when reporting purchasing power parity (PPP) data.

References

Arrhenius, O., 1921, "Species and area," *Journal of Ecology*, 9(1): 96–99.
Baltagi, B. H., 2001, *Econometric Analysis of Panel Data*, 2nd edn, New York: John Wiley and Sons.
Blundell, R. and Bond, S., 1998, "Initial conditions and moment restrictions in dynamic panel data models," *Journal of Econometrics*, 87(1): 116–143.
Cardinale, B. J., Ives, A. R. and Inchausti, P., 2004, "Effects of species diversity on the primary productivity of ecosystems: Extending our spatial and temporal scale of inference," *Oikos*, 104(3), 436–450.
Coleman, B. D., Mares, M. A., Willig, M. R. and Hsieh, Y. H., 1982, "Randomness, area, and species richness," *Ecology*, 63(4): 1121–1133.
Di Falco, S. and Chavas, J. P., 2008, "Rainfall shocks, resilience, and the effects of crop biodiversity on agroecosystem productivity," *Land Economics*, 84(1): 83–96.
Extended Penn World Table 3.0. URL: http://homepage.newschool.edu/~foleyd/epwt/ (Accessed: 21 May 2010)
FAOSTAT. URL: http://faostat.fao.org/site/377/default.aspx#ancor (Accessed: 25 May 2010).
Food and Agriculture Organization (FAO), 2001, *Global Forest Resources Assessment 2000*, FAO Forestry Paper 140.
Gaston, K. J., 2000, "Global patterns in biodiversity," *Nature*, 405: 220–227.
Kitamura, Y., 2005, *Paneru deta bunseki* (Panel Data Analysis), Tokyo: Iwanami Shoten (in Japanese).
McGuinness, K. A., 1984, "Equations and explanations in the study of species-area curves," *Biological Reviews of the Cambridge Philosophical Society*, 59(3): 423–440.
Millennium Ecosystem Assessment (MA), 2005a, *Ecosystems and Human Well-Being, Synthesis*, Washington, DC: Island Press.
——, 2005b, *Ecosystems and Human Well-Being, Volume 1: Current State and Trend*, Washington, DC: Island Press.
Richmond, A., Kaufmann, R. K. and Myneni, R. B., 2007, "Valuing ecosystem services: A shadow price for net primary production," *Ecological Economics*, 64(2): 454–462.
TEEB, 2010, *The Economics of Ecosystems and Biodiversity—Ecological and Economic Foundations,* P. Kumar, (ed.) London: Earthscan.
Tilman D., Polasky, S. and Lehman, C., 2005, "Diversity, productivity and temporal stability in the economies of humans and nature," *Journal of Environmental Economics and Management*, 49(3): 406–426.
Tsutsui, J., Hirai, H., Akiyoshi, M., Mizuochi, M., Sakamoto, K. and Hukuda, N., 2007, *Stata de Keiryo Keizaigaku Nyumon* (Introductory econometrics with the Stata), Kyoto: Mineruva Shobo (in Japanese).
Turner, J. R. G., Lennon, J. J. and Lawrenson, J. A., 1988, "British bird species distributions and the energy theory," *Nature*, 335: 539–541.
Tyndall Centre for Climate Change Research. URL: http://www.cru.uea.ac.uk/~timm/cty/obs/TYN_CY_1_1.html (Accessed: 1 June 2010).
World Development Indicators. URL: http://data.worldbank.org/indicator/EG.USE.COMM.KT.OE (Accessed: 22 May 2010).
Wright, D. H., 1983, "Species-energy theory: An extension of species-area theory," *Oikos*, 41(3): 496–506.

Yamamoto, T., 1992, *Jikeiretsu Bunseki to sono Keizai Bunseki heno Oyo* (Time-series analysis and application to economic analysis thereof), Okurasho Zaisei Kinyu Kenkyujo, Financial Review, June (in Japanese).

Part III

Economic instruments for ecosystem conservation

7 Towards the establishment of a Payments for Ecosystem Services policy framework

Seiji Ikkatai

Overview of Payments for Ecosystem Services

The conservation of nature has conventionally been carried out through regulatory measures such as: designation of protected areas; limitations on exploitation of land, forests, or wildlife; regulations on the capture and trade of rare species; or through tax incentives such as property tax exemptions for portions of land located within important conservation zones of national parks or other protected areas. However, these measures alone cannot generate resources for the long-term maintenance of protected areas or species. As a result, biodiversity loss and degradation have been exacerbated beyond protected areas in the wider urban and rural ecosystems.

In addition to these conventional nature conservation policies, alternative policies for the conservation and sustainable use of biodiversity are being developed internationally and domestically, based on an economic instrument called Payments for Ecosystem Services (PES).

Ecosystem services are defined as "the conditions and processes through which natural ecosystems, and the species that make them up, sustain, and fulfill human life" (Daily 1997: 3). Thus PES can be understood as a means to integrate ecosystem services into economic activities through proper assessment and recognition of both the tangible and intangible benefits that humans obtain from nature. In other words, until now people have been free-riding on ecosystem services, but through economic mechanisms to pay for their use based on the Beneficiary Pays Principle (BPP), a socioeconomic system can be established for the sustainable use of natural resources by feeding back financial resources to their protectors and managers such as farmers.

The flow of ecosystem services can be categorized into three types: (1) from upstream to downstream (water use within a river basin); (2) from rural to urban (products and services from rural areas that are provided to and consumed within urban areas); and (3) between North and South (global transfer of ecosystem services between developed and developing countries through international trade). Within Japan, cases of PES-like measures illustrating these flows of ecosystem services include in order: (1) the Forest Environment Tax or Water Source Tax (fees added to water bills for the maintenance of catchment forests), and payments from downstream companies relying on their water supply to owners of upstream forests or paddy fields; and (2) the Local Allocation Tax and government subsidies applied to the maintenance of forests and natural parks. As in these examples, it is possible to mandate the payments for ecosystem services through laws and taxation, but this will primarily require a consensus on cost sharing to be built among beneficiaries (through contracts or committee meetings).

Under PES schemes, the beneficiaries of ecosystem services – that is those who should make the payments – are diverse. They can roughly be grouped into: (1) the general public; (2) local inhabitants; (3) ecosystem service consumers; and (4) the private sector. Concrete PES mechanisms can take the form of direct payments from ecosystem services beneficiaries to providers, taxes or fees, or a price premium for certified products.

PES has already been adopted as a national strategy by countries like Costa Rica, but is still far from becoming a globally standardized policy. Therefore, various research activities on concrete policy approaches to PES are being conducted across the world, including Japan. Diverse possibilities on policies and actions for PES were discussed at the CBD-COP10 held in Nagoya City in October 2010; clauses on PES policy implementation have been included in the 2020 Aichi Targets and related decisions and policy documents.

When considering PES, climate change mitigation policies are often referred to as an antecedent, especially with regard to the MRV[1] of carbon emission reductions and its possible linkages to the development of ecosystem services trading approaches (payment mechanisms). The current theories and policy measures for PES are still underdeveloped, but they are expected to evolve into internationally recognized strategies for the internalization of external costs for ecosystem services in order to achieve the conservation and sustainable use of biodiversity.

Cases of PES-like measures in Japan

Japan possesses a rich and diverse natural heritage, thanks to its long coastline and a larger proportion of forested land in comparison to other developed countries. Traditionally, Japanese society had developed in a sustainable manner centered around rice farming which relies on an abundant water supply from its forests. In particular, during the Edo era, Japan is thought to have established a rare, ecologically sound society under a strong national seclusion policy. Administrative units were set on the basis of natural landscape boundaries, and as energy use was limited to renewable sources, society was able to preserve its natural environment prevent pollution, and manage its waste effectively.

However, since the Meiji Restoration when Japan opened its doors to the world and underwent rapid development with the introduction of Western culture, the advanced environmental know-how was forgotten, leading to many serious environmental problems. Natural ecosystems were not spared the consequences, with forest degradation, farmland abandonment, destruction of natural coastlines for land reclamation, and urban heat islands being examples of their demise.

Although circumstances have changed in Japan, parts of the historical knowledge on the conservation of nature through PES-like arrangements have been inherited by today's society. It is thus worth exploring the overall picture of Japanese cases of PES-like measures. No strict limits will be set to define what PES-like measures encompass, but rather, a broad introduction will be given to current measures that allow financial flows towards the provision of ecosystem services.

Historical case studies

According to a report on the history of downstream participation in the establishment of water catchment forests (Kumazaki 1981), the people of Mizuno village in Kubiki District, Echigo Province (today's Niigata Prefecture), submitted a request for approval in 1784 to start charcoal production in the community forest, but they were met by strong protests from the 24 villages situated further downstream. The downstream villagers were concerned that "forest exploitation would lead to water shortages due to accelerated thaw in spring, and also create a higher risk of sediment discharge in cases of rain" (Kumazaki 1981). An agreement was reached for Mizuno village to cancel the forest-clearing and charcoal-production plans, by receiving in exchange "50 Ryo [a past currency] as a first payment, and a yearly compensation of 150 kg of rice" (Kumazaki 1981). Furthermore, it was stipulated in their agreement that "when forest overgrowth causes deer and boars to ravage the adjacent farmland, tree felling will be conducted under the witness of both upstream and downstream villages, following the upstream farmlands' maintenance needs but respecting the limit of potential impacts to water provision" (Kumazaki 1981).

This case can be seen as a payment by the downstream communities to the upstream communities for the preservation of a stable water supply, as a forest ecosystem service, which is needed for rice cultivation. Additionally, it can be observed that the agreement between the upstream and downstream communities included surprisingly detailed arrangements for conducting limited felling in the upstream forests in case the cancellation of exploitation led to overgrowth and the invasion of boars and deer into farmlands. It can be expected that such agreements that include PES-like measures between upstream and downstream communities were not unique to the case of the Kubiki District, but existed in diverse forms across Japan.

Later in history, in 1900, a case of district-managed forest plantation can be found in the Inukami District of the Shiga Prefecture. In this district, there are two rivers – the Seri River and the Inukami River – which used to feed over 3,960 hectares of rice paddies, but which were short of water, leading to frequent conflicts during droughts. On the other hand, the area was also prone to floods when there were heavy rains. The main cause of these issues was the degradation of water-catchment forests due to the exploitation by the Hikone domain, illegal logging within community forests, and frequent field-burning activities. It was thus decided that the district government would borrow forested land in two areas – Ojigahata, upstream from the Inukami River, and Unzen, upstream of Seri River – in order to conduct afforestation. However, this received strong opposition from landowners because local inhabitants would not be able to make a living if limits were placed on the extraction of material for coal from forests, and also because afforestation activities had already been initiated in some areas. As a result of negotiations, the Inukami District government was able to obtain land-use rights in 1900 for 4.95 hectares of land in each area with a revenue-sharing contract of 100 years, and a revenue-distribution agreement of 90 percent for the district government and 10 percent for the landowner. Additionally, a land-rental contract was signed in 1902 for 1.15 Japanese yen (JPY) per hectare and for 1.50 JPY per hectare in 1911. Since obtaining land-use rights, the district government applied a total of 67,000 JPY up to 1907 (JPY value of the time) for afforestation activities.

This record shows a case where the district government bore the costs of preventing upstream forest exploitation and for afforestation, and does not necessarily imply a direct payment between upstream and downstream communities. However, the measures benefit more than 3,960 hectares of downstream rice paddies within the district, so there must have been negotiations and agreements among the beneficiaries regarding the financial burden of the district government. Also, many similar cases can be seen in community-managed forests of Koka District in Shiga Prefecture, or of Yame District in Fukuoka Prefecture, as well as in the management of revenue-shared forests of the Meiji irrigation channel's soil improvement areas that are situated upstream of the Yahagi River. These management measures were introduced with the aim of preventing forest degradation and ensuring aquifer recharge, but the fundamental reason for the revenue-sharing agreements and land-rental fee payments to the landowners was because the profitability of the wood market was considerably higher than in recent times.

Later, in the Showa Era, there have been cases of forest management by prefectural governments in partnership with electric power companies. For example, in Tochigi Prefecture, a prefectural Catchment Forest Management Fund was established in 1979, and forest maintenance activities were initiated within the prefecture. This fund was established with donations from the electric power company and transfers from the Enterprise Bureau's budget of the prefectural government. The revenues of the fund were used as subsidies for tree nurseries maintained by forest owners. This kind of measure was conducted in many parts of the country in response to lagging forest maintenance by private owners due to an insufficiency of government support for forest-nurturing activities, as well as to the decline of domestic forestry industries. This period was also marked by a rapid increase in electricity demand with the economic growth of the postwar period, and thus the maintenance of water catchment forests was given a top priority by electric power companies for hydroelectric power generation. It is thus possible to interpret the electric power companies' support towards forest nurturing activities by forest owners as a payment for ecosystem services and their maintenance.

Case of forest environment taxes by local governments

In the context of postwar economic growth and trade liberalization, the import of low-cost foreign timber increased considerably, causing the profitability of domestic forestry industries to plummet. As a result, the degradation and lack of maintenance of forest nurseries were accentuated, especially among privately owned forests. In such a context, increased attention to environmental problems as well as the enactment of the government's Decentralization Law in 2000 boosted the introduction of Forest Environment Taxes among local governments. The main aim of this prefectural tax is to secure funds for forest maintenance through widespread burden-sharing among the citizens who benefit from the water storage and other environmental functions of forests. The first to implement this system was Kochi Prefecture, which established a prefectural ordinance to introduce the Forest Environment Tax in 2003, followed by Okayama Prefecture, Tottori Prefecture, and many others. By April 2009, there were 30 prefectural governments and one city government which had introduced this tax.

Although the uses of tax revenues are diverse, these systems can be subdivided into three categories according to the particular taxation approach.

The first category follows Kochi Prefecture's approach, which consists of placing a fixed 500 JPY surcharge on the Individual Resident Tax and Corporate Tax. The second category follows Okayama Prefecture's example, which imposes a fixed surcharge on the Individual Resident Tax and imposes an additional percentage on the Corporate Tax. In this second approach, the individual surcharge can range from Shizuoka Prefecture's 400 JPY to the Fukushima and Iwate prefectures' 1,000 JPY. As for the Corporate Tax surcharge rate, it can range from the 5 percent of the Standard Tax imposed by Shizuoka and 20 other prefectures, up to the 11 percent of the Standard Tax imposed by Shiga Prefecture. The third category follows Kanagawa Prefecture's approach, which imposes 300 JPY per year and a rate of 0.025 percent of income on individual residents, and no additional taxes for corporations. In Kochi Prefecture, which was the first to introduce this system, two different approaches had initially been considered, namely that of imposing an additional tax percentage based on water consumption levels, and that of imposing a surcharge on the resident tax; the latter was finally selected due to its simplicity. Also, as this surcharge taxation approach does not limit the uses of its income, many local governments apply the tax income to their newly established Forest Environment Fund in order to support maintenance reinforcement such as forest-thinning activities, diversification of tree species, nurturing of rural, coastal, and school-owned forests, and the securing of carbon sequestration sources, as well as to support social reinforcement such as residents' participation in forest maintenance, environmental education, revitalization of forestry, and the fostering of successors for the local forestry industry. In most prefectures, these measures are conducted, not by the forest owners, but by a third party such as the prefectural government or the local forestry associations, based on mutual agreement between the owners and the prefectural government to place a temporary limit on the property rights of the forest.

Thus, since the introduction of the Forest Environment Tax in Kochi Prefecture in 2003, Japan has been seeing a rapid spread of PES-like measures for water supply in the form of tax surcharge payments by the citizens who are beneficiaries of the forests' water-retention function. However, it remains a challenge to ensure that the taxation rates and forest maintenance activities, as well as educational activities, meet the level of ecosystem services provided by the forests.

Other cases of funds linking upstream and downstream communities

In Japan, prior to the establishment of the aforementioned Forest Environment Tax, local governments have taken the initiative to share costs and benefits between upstream and downstream communities through funds developed from surcharges on water-usage bills.

For example, Toyota City set up the Toyota City Water Source Preservation Fund in 1994 with the aim of securing a "safe and tasty" tap water supply through environmental maintenance, to ensure water storage in aquifers, and water-quality preservation. More precisely, the city government added a surcharge of one JPY per cubic meter onto the water bills (collecting 46 million JPY per year) which was fed into the Preservation Fund. From the year 2000 onwards, an area of over two hectares has been selected as a watershed preservation forest among the privately owned forests, and an annual sum of 16 to 29 million JPY from the city's fund has been applied to their maintenance, as well as to support the installation of high-degree septic systems in upstream households. Similar cases can be seen in other regions such as the Toyo River Watershed Fund in the Aichi Prefecture or the Fukuoka Prefectural Water Service and Watershed Preservation Fund. Fund systems that link the upstream and downstream communities within watersheds were established in 44 local governments across the country by 2009.

Cases of direct payments in agriculture

As in the case of forestry, the profitability of agriculture in Japan was hard hit by the postwar economic growth and the increase in low-cost foreign food imports resulting from trade liberalization. Consequently, agricultural land saw rapid degradation due to its abandonment, especially in hilly and mountainous areas where the aging of the local population poses an additional obstacle to generating profits.

Therefore, in 1998 the national government set up a direct supporting system for mountainous regions as a means to secure the agricultural functions and policies within hilly areas which are suffering from abandonment and degradation. Targeting the particularly steep fields of selected villages, the government has been providing financial support, per surface area, amounting to the difference in production costs compared to normal farmlands, with an upper limit of one million JPY per recipient. The total spending for this support system exceeds an annual amount of 50 billion JPY. When looking at the records for the financial year 2009, 1,008 municipalities have received financial support, amounting to a surface area of 664,000 hectares, and 28,309 communities. Per community, the average number of members is 23 people and the average allocated subsidy amounts to 1.82 million JPY, which is equivalent to 80,000 JPY per person. The Ministry of Agriculture, Forestry, and Fisheries (MAFF) intends to prevent farmland abandonment and to foster the multiple functions of agriculture towards securing long-term productivity and the vitality of farming communities.

Additionally, with the shift in political power to the Democratic Party in 2009, the national budget for the farmers' income compensation scheme was secured in line with the Democratic Party's manifesto. The target of this scheme in 2010 was limited to rice farmers but was extended to include other cereals in 2011. This scheme aims to maintain the agricultural function and to increase Japan's food sufficiency through the stabilization of agricultural business and domestic productivity. For instance, for rice farmers, this scheme provides a direct payment of 15,000 JPY per 1,000 square meters of farmland. From 2011, a government budget was secured for a new direct payment scheme for environmentally sound agricultural practices which builds upon past farmland, water, and environmental preservation measures. This new scheme provides direct payments to farmers who apply agricultural practices that are effective for climate change mitigation or biodiversity conservation at a rate of 4,000 JPY per 1,000 square meters of farmland.

There are also examples of direct payment schemes conducted by local governments. In Osaki City, Miyagi Prefecture, a support scheme for organic farming in the Kabukuri swamp and for winter-flooding of rice paddies is provided based on regional agricultural subsidies. It aims to back up farmers willing to introduce winter-flooding of paddy fields and organic farming in order to provide biodiversity-friendly wintering grounds for wild geese. Under this scheme, farmers are subsidized with 8,000 JPY per 1,000 square meter, and the rice produced from these paddies is branded and sold with an environmental premium. Also, in Shiga Prefecture which aims to reduce the environmental impacts of agriculture, the Environmental Agriculture Direct Payment Scheme has been established to encourage a reduction in the use of chemical fertilizers and to support farmers introducing technologies that reduce their impact on the environment and that satisfy a certain number of criteria (970 hectares subsidized by 1996).

Cases of PES-like schemes in urban areas

The enhancement of the quality of life by ecosystem services in urban areas is not limited to therapeutic effects from water, greenery, and wildlife, but also other benefits such as heat-island mitigation.

Among the PES-like measures implemented by the national government, the Productive Green Space Promotion Scheme is an example. This scheme aims to reduce the financial burden of maintaining green spaces which measure over 500 square meters by introducing tax exemptions on property taxes, and by allowing a postponement of the payment of land inheritance taxes, and it contributes, to a certain extent, to the conservation of green spaces within urban areas. Among schemes implemented by local governments, an example is the Protected Tree designation in the Nerima ward of Tokyo. This scheme aims to preserve trees measuring over 1.5 meters in height and over 50 centimeters in diameter, or forested areas of over 1,000 square meters. The selected trees or areas can receive subsidies on maintenance costs and on the costs of liability insurance. Similar systems can be found in numerous municipalities, but since the majority is based on voluntary actions by the landowner, there is no guarantee of continued maintenance in cases such as a change of ownership.

Cases of catchment forest conservation by private companies

In Japan, there have traditionally been many cases of forest maintenance and conservation funded by companies relying on forests for water supply, such as beer breweries. For instance, Asahi Breweries already held the notion in 1941 that "forest protection leads to business protection." It began from the purchase of forests in Hiroshima by the predecessor company Dainihon Beer which sought to secure oriental oak trees for their bark, which is used as a substitute for cork. Asahi later inherited the forest maintenance role when the company was separated, and is now managing it as Asahi's forest. Asahi has conducted environmentally conscious forest management until today, and in 2001 became the first in Japan to obtain the internationally recognized FSC (Forest Stewardship Council) certification. Asahi now owns 2,165 hectares of forests in 15 areas of varying sizes within Shobara City and Miyoshi City.

Kirin Holdings and Suntory Holdings have been carrying out similar activities in collaboration with local governments since 2005 and 2003, respectively. Their activities for watershed protection involve afforestation through their "movement for the protection of the bounty of water" and of the "forests of natural springs." More recently, these forest and watershed conservation activities have been spreading among other businesses such as commercial companies, vehicle manufacturers, department stores, and semiconductor vendors.

Other recent trends

As previously introduced, there have been diverse payments for forest maintenance and conservation centered on their water storage functions but, in the context of aggravating climate change issues, there have been increasing examples of payments for the carbon storage function of forests. For example, as a carbon dioxide sequestration measure based on the Kyoto Protocol, the national government set up in 2002 a 10 Year Policy Framework for Forest Carbon Storage, for the promotion of forest nurturing and the appropriate maintenance of private forest reserves through forest thinning and additional reserve designation. The common cost-sharing proportions are 50 percent by national government subsidies, 20 percent by prefectural government subsidies, and 30 percent by the forest owner. With the aim of promoting voluntary activities in the private sector, and with the establishment of the national guidelines on carbon offsetting in 2008, new schemes have been developed among local governments for the crediting of carbon storage through forest maintenance, and for the trade of offset credits among private companies. These schemes can also be seen as PES-like measures that focus on one of the diverse ecosystem services of forests.

Towards the development of a PES policy framework

Challenges for policy development from domestic PES-like schemes

An overview of domestic PES-like schemes has been provided. However, these schemes have not been developed and implemented with the purpose of paying for ecosystem services. Rather, they have focused on other concrete needs such as to secure sources for water provision or electricity generation, to find solutions to the degradation of forests or farmlands, to prevent natural disasters such as floods, or, more broadly, as part of private companies' social contribution. Therefore, the levels and mechanisms of cost sharing are still uneven.

From the perspective of PES policy development, there are many challenges that would need to be overcome.

First, the targets of these domestic cases are almost entirely limited to parts of forest, farmland, and urban ecosystems. For a PES policy, other ecosystems such as coastal, marine, and wetland ecosystems would be of high importance, but cases of payments for services from these ecosystems have yet to be identified.

Among the current examples of PES-like schemes, the closest to a PES scheme would be the local governments' Forest Environment Tax, which taxes beneficiaries of forest ecosystem services. However, this system targets the water storage function of the forest, and does not necessarily link payments from beneficiaries to other forest ecosystem services such as carbon storage or biodiversity maintenance. Even when focusing on locally beneficial ecosystem services such as water storage, there has been no evaluation to determine whether the levels of taxation and the uses of tax incomes meet the value of ecosystem services acquired. Furthermore, regarding the carbon storage functions, the beneficiaries are not limited to prefectural residents, but this taxation system is confined within the prefecture. On the other hand, the system which has the potential to address this issue is the carbon absorption crediting system, which allows credits to be sold outside the prefecture to stakeholders such as private companies. Therefore, PES schemes for forest ecosystem services could be composed of a combination of different payment schemes, depending on the ecosystem service in question.

Second, with regard to PES from farmlands, there has been a stronger focus on hilly and mountainous areas, and schemes to conserve farmlands broadly have been limited up to the present time. If the Income Compensation System for farmers is put into effect from the financial year 2011, the recipients of subsidies will expand largely beyond rice producers. However, as previously mentioned this system includes many policy objectives other than the preservation of farm-land ecosystem services, such as the stabilization of agricultural production and securing domestic food sufficiency. There is thus a need for further investigation on whether the payment standards actually contribute to the maintenance of ecosystem services.

Third, regarding urban ecosystems, there are a few PES-like schemes based on national and local government initiatives, but as they rely on landowners' voluntary actions, their effects on the conservation and long-term stable supply of ecosystem services may be limited.

Fourth, regarding the future establishment of a PES policy framework, local governments have been promoting PES-like schemes centered on Forest Environment Taxes, and PES-like schemes of direct payments to farmers have been attempted. However, it must be pointed out that a national level PES scheme that encompasses all of the diverse ecosystems has not been envisioned nor positioned vis-à-vis national policies, and the basic pathway towards its introduction has not been defined. Also, within national and local government schemes, most PES-like payments have taken the form of subsidies, with the exception of the Forest Environment Tax. Subsidies generally rely heavily on the financial situation of governments, which are expected to become increasingly severe in the future, and thus the establishment of a stable payment system based on the current approach would be unlikely.

Challenges towards the establishment of PES policy schemes in Japan

In order to establish a new and important scheme such as PES, there is a strong need for its basic principles and positions within national policies to be understood and shared by the citizens. This is because a central principle on what costs the citizens must bear for what kind of services is needed as an axis of the nation's socioeconomic system.

During postwar Japan's rapid economic growth, national and local governments played an important role in the development of the country through the accumulation of social capital. Such social capital includes road infrastructure, ports, airports, public transport infrastructure, and electricity sources, many of which were centered on construction work that led to direct economic returns. However, the maintenance of ecosystem services derived from natural capital including urban, farmland, forest, coastal, and marine and wetland ecosystems – indispensable to human survival – have not been considered from the perspective of the development of social capital. Therefore, this natural capital which is not sufficiently internalized in the market economy has gradually become degraded, leading to serious challenges in the establishment of socioeconomic sustainability.

In this context, it is necessary to position clearly the maintenance and accumulation of natural capital (urban, farmland, forest, coastal, marine, and wetland ecosystems) as an important component of social capital within national policies. During this process, the challenges and shortcomings of current PES-like schemes will become apparent, both in the target areas and in the level and stability of the payments. Bearing this in mind, decisions will be made possible on whether existing PES-like schemes should be directly applied in policies, whether they should be improved, or whether a completely new scheme needs to be created.

For the design of a PES scheme, the challenge will be to identify the source of funds for the payments. There are many possible approaches including general taxes such as income tax or VAT, taxes focusing on environmental burden such as an environmental tax, surcharges with a defined purpose, or direct payment by consumers through pricing systems. Ideally, beneficiaries of each ecosystem service should bear the costs according to their level of dependence on the service, but since market values cannot always be attributed to ecosystem services, it would not be realistic to attempt to create such schemes for all ecosystem services. Therefore, the kind of cost-bearing mechanism appropriate for each type of ecosystem service should be considered holistically, taking into account factors such as equity, efficiency, and administrative costs. In this regard, the combination of numerous schemes for different kinds of ecosystem services – as in the case of the Forest Environment Tax and carbon storage credits for the ecosystem services of forests – show a way forward in designing a PES scheme in Japan.

However, PES schemes are not meant to be a unique solution for the maintenance and enhancement of ecosystem services. Therefore, PES schemes need to be managed as part of a broader policy agenda for the securing and improvement of natural capital, and in parallel with the creation of a no-net-loss policy that should be applied to preexisting zoning and development regulations. Therefore, it is critical to continue further considerations of reforms in zoning regulations and of the establishment of no-net-loss policies.

Ultimately, the amount of payments should be determined based on the valuation of each ecosystem service, but given the difficulties in monetizing these values it would not be possible to rely on a conventional economic model where the amount of payments determines the amount of services secured. Instead, for the time being, the PES scheme would need to be positioned as a supporting mechanism to secure and enhance the nation's natural capital, based on citizens' consent.

From such a perspective, the "Green Infrastructure" policy put forward by the European Commission's Environment Directorate-General has the potential to serve as an overarching framework when combined with a PES scheme that allows the generation of financial resources for the maintenance of natural capital. This policy aims to secure a healthy natural capital based on the understanding that human society's capital including railways, roads, and water works all derive from diverse ecosystem services. Furthermore, this Green Infrastructure policy should attempt to integrate fragmented policy measures relating to ecosystem management under one umbrella, and to secure financial backing for their implementation. Examples of such policies include nature and biodiversity conservation, land-use, water management, marine and coastal management, transport and energy, climate change, urban development, resource-efficiency improvement, and environmental-impact assessment policies. As a means to fulfill the goals of the Green Infrastructure policy, focus is placed on the maintenance and restoration of regional ecosystems and of regional water and carbon cycles. This policy, nevertheless, is still under development; thus, diverse attempts at national and regional levels are expected to contribute to its further evolution.

Acknowledgements

The author would like to thank Ms Sana Okayasu, Associate Researcher at the Institute for Global Environmental Studies, for her translation of the original Japanese manuscript into English.

Note

1 MRV is Measurement, Reporting, and Verification of emission reductions under climate change policies.

References

Daily, G., 1997, *Nature's Services: Societal Dependence on Natural Ecosystems*, Covelo, CA: Island Press.
Kumazaki, M., 1981, "Suigenrin ni okeru karyu sanka no keifu (Genealogy of downstream participation in water source forest management)," *Mizukagaku* (Water Science), 3: 1–24 (in Japanese).

8 Impact assessment of sustainable forest use policy with the economic model

Satoshi Kojima and Kei Kabaya

Introduction

The concept of ecosystem services covers a very wide range of services, not only those of direct use but also various regulating services such as flood regulation and atmospheric-composition stabilization, as pointed out by the Millennium Ecosystem Assessment (2005). For example, global warming can be interpreted as a problem of sustainable use of an ecosystem service, that is, atmospheric-composition stabilization. Conservation of biodiversity, another important global environmental issue, can also be interpreted as a problem of sustainable use of ecosystem services, because either biodiversity itself provides ecosystem services or conservation of biodiversity is essential to produce some ecosystem services. Along this line of argument, the concept of sustainable development proposed by the Brundtland Report (World Commission on Environment and Development 1987) can be redefined as the alleviation of poverty (in the sense of poverty as a lack of opportunity to meet basic needs) through utilizing ecosystem services in a sustainable manner.

Despite the political importance of these matters, there are not many existing studies analyzing the sustainable use of ecosystem services in a quantitative manner. This chapter presents a study which conducted a quantitative impact assessment of policy aimed at the sustainable use of ecosystem services, based on a general equilibrium model which can take account of interactions among various economic actors such as households and firms.

Here, we briefly explain economic models and general equilibrium models. In economic models, an economic system is assumed to consist of households, producers, and the government. Households earn income by providing factors to the production sectors, and they expend this income either for purchasing various commodities or for saving. Producers, consisting of various industries, produce goods and services using factor inputs (labor and capital) purchased from households and intermediate inputs purchased from other producers. The government implements policies. If the analyzed policies are economic instruments such as taxes and subsidies, the government collects these taxes and pays subsidies. In contrast to partial equilibrium models which focus on specific segments of an economic system (say, agriculture), general equilibrium models deal with all components of an economic system in a comprehensive manner. General equilibrium models are suitable for analyzing policies which have economy-wide impacts through interlinkages between various sectors or indirect impacts through households' decisions on consumption and labor supply. General equilibrium models for empirical analysis based on actual economic (and other) data are called computable general equilibrium (CGE) models. CGE models are widely used for quantitative policy assessment in various fields, in particular climate-policy studies.

One of the pioneering works applying general equilibrium models to sustainable use of ecosystem services is Washida (1999), which introduces ecosystem services into the DICE model (Nordhaus 1993), a Ramsey-type dynamic general equilibrium model. Washida assumes that ecosystem services can substitute for consumption of commodities in the utility function of the DICE model, and analyses the relationship between people's economic valuation of ecosystem services provided by closed forest ecosystems and dynamic paths of use and conservation of these services. The DICE model is a 1-region 1-sector model covering the whole world and Washida (1999) undertakes numerical simulations of a theoretical model dealing with the impacts of the elasticity of substitution between consumption and ecosystem service. This represents households' valuation of the scarcity of ecosystem services on the use and conservation of the services, rather than being an empirical policy assessment of specific regions. Furthermore, the model reflects only afforestation, but does not consider natural growth of forests to explain forest-stock accumulation.

Sohngen and Mendelsohn (2003) combine the DICE model with a global forest model and discuss the necessity of coordinated actions between GHG emissions mitigation from an increased carbon sequestration capacity of forests and other mitigation measures such as energy saving. This study focuses on long-term carbon price paths satisfying carbon constraints until the year 2100, and on the amount of sequestrated carbon, but not on the sustainable forest use perspective. There are more empirical studies such as that of Ahammed and Mi (2005) which introduces the carbon sequestration function of forests into a multiregional multisectoral CGE model. These studies mainly focus on GHG mitigation by forests, but not on sustainable use of forest ecosystem services in a direct manner.

Hamilton (1997) conducts an environmental impact assessment of four Indonesian scenarios with respect to (1) the ratio between logging in plantation forests and those in natural forests, and (2) economic growth rates, assuming that sustainable logging technologies will be introduced more widely in natural forests. The study is unique in combining a macroeconomic model and a relatively detailed environmental-impact assessment, which takes account of chemical fertilizer and agrochemical use, as well as soil runoff of the forestry sector, given exogenous growth of final demands. This study is, however, not an economic impact assessment of sustainable forest use.

With the above research gap in mind, this chapter presents an impact assessment of sustainable-forest-use policy in Indonesia using a policy-impact assessment model, combining a CGE model and a forest-stock model reflecting natural forest growth. Indonesia is endowed with very rich tropical rain forests, which are globally very precious in the same way as those in the Amazonian watershed and Congo River basin. The rapid increase in commercial logging under the Suharto administration, however, has meant the loss of a significant proportion of forests, estimated at 40 percent (Engel and Palmer 2008) over the 50 years between 1950 and 2000. Furthermore, the current logging method is pointed out as being unsustainable, based on the fact that it has caused various environmental problems and it has failed to maintain the logging rate (Hamilton 1997). It has been estimated that lowland forests in Sumatra and Kalimantan will disappear within ten years or so (Forest Watch Indonesia and Global Forest Watch 2002). In Indonesia, more than 6 percent of the employed population is engaged in forest-related industries (forestry, lumber industry, and the paper and pulp industry), and forest resources play a very important economic role. Indonesian tropical rainforest is blessed with very diverse fauna and flora; it is estimated that Indonesia possesses 10 percent of global plant species, 12 percent of mammal species, 16 percent of reptile species, and 17 percent of avian species (Hamilton 1997). Moreover, it is estimated that Indonesian fauna and flora fixes 3.5 billion tonnes of carbon (Forest Watch Indonesia and Global Forest Watch 2002). Thus, Indonesian sustainable forest use is a very important political objective in promoting sustainable development, not only for Indonesia but also for the world.

In this chapter, we formulate the business-as-usual (BAU) scenario reflecting the current forest-use situations and the sustainable forest-use (SFU) scenarios aimed at forest-stock conservation. We assess the economic impacts of SFU by comparing the simulation results of the SFU and the BAU scenarios using a multisectoral Ramsey-type dynamic CGE model with forestry and lumber sectors. This chapter also covers policy instruments designed to transform ecosystem services to economic benefits, such as the reduced emissions from deforestation and degradation (REDD), which draws wide attention in the international negotiation process regarding climate change, and the payment for ecosystem services (PES).

Policy impact assessment model

The policy impact assessment model employed in this chapter consists of a forest-stock model which describes policy scenarios, and an economic assessment model based on a multisectoral Ramsey-type dynamic CGE model which assesses the economic impacts of the policy scenarios.

The forest-stock model – a simple Gordon–Schaeffer type model representing forest growth – describes dynamic paths of logging volume, forest stock, and forest area under each policy scenario. The input variables from the forest-stock model to the economic assessment model are logging volume, change in efforts per unit of logging volume due to forest-stock depletion, and REDD credits for increased carbon fixation due to forest-stock conservation. The latter two variables are obtained as functions of forest stock.

The economic assessment model reflects these inputs as follows. The percentage change in logging volume is input as a natural resource input to the forestry sector. Similarly, the reciprocal of the percentage change of efforts per unit of logging volume is input as the total factor productivity (TFP) of the forestry sector. The total quantity of REDD credits is modeled as lump-sum transfers from the rest of the world (foreign countries) to the government.

The forest-stock model and the economic assessment model are explained below.

Forest-stock model

Outline of the model

The forest stock model in this analysis is a dynamic model integrating natural forest growth, selective logging, clear cutting, illegal logging, land conversion, forest fire, and plantation at the end of the year. Forest growth in the area of original forest, selective logging, and plantation can be factors for increasing the forest stock, assuming that their growth rates depend on each forest's density. On the contrary, selective logging itself, clear cutting, illegal logging, land conversion, and forest fires can contribute to the loss of forest stock, and it is presumed that forest areas themselves are decreased, except in the case of selective logging.

First, we premise that natural growth will draw the Gordon–Schaeffer type logistic curve as the following model (cf. Chopra and Kumar 2004):

$$V_{t+1} = V_t + \varphi(D_t)V_t \left[1 - \frac{D_t}{D_{\max}}\right],$$

where V denotes forest stock, subscript t is year (throughout our analysis), and D, $D_{\max}$, and $\phi(D_t)$ indicate forest density, maximum density, and the intrinsic natural growth rate dependent on forest density, respectively. Here, $\phi(D_t)$ is assumed to draw an inversely proportional curve in response to D_t and to take the value of 0.03 around the maximum figure of $D/D_{\max}$, 0.05 around the middle, and 0.10 around the minimum thereof, with reference to Goodland et al. (1991).

Selective logging decreases the forest stock, depending on its logging rate; however, the rest of the forest stock in those areas is expected to grow faster than the original one did. Meanwhile, clear cutting, illegal logging, land conversion, and forest fires are diminishing forest areas, thereby making the forest stock in those areas disappear completely. As it will be difficult to see natural regrowth in the areas where vegetation has been totally removed, it is supposed that no regrowth will occur in these areas. As for the forest stock in the plantation area, we assume that it will grow at an accelerated pace in spite of its low density at the beginning. Based on the above assumptions, the forest area and stock at time $t + 1$ are modeled as follows:

$$A_{t+1} = A_{o,t} - A_{c,t} + A_{p,t}$$

$$V_{t+1} = \left(V_{o,t} - \frac{V_{s,t}}{\omega} - V_{c,t} - V_{i,t} - V_{cv,t} + V_{f,t} \right)$$

$$+ \varphi_{o,t} \left(V_{o,t} - \frac{V_{s,t}}{\omega} - V_{c,t} - V_{i,t} - V_{cv,t} + V_{f,t} \right) \left[1 - \left(\frac{D_{o,t}}{D_{\max}} \right) \right]$$

$$+ (1 - \omega) \left(\frac{V_{s,t}}{\omega} \right) + \varphi_{s,t} (1 - \omega) \left(\frac{V_{s,t}}{\omega} \right) \left[1 - \left(\frac{D_{s,t}}{D_{\max}} \right) \right]$$

$$+ V_{p,t} + \varphi_{p,t} V_{p,t} \left[1 - \left(\frac{D_{p,t}}{D_{\max}} \right) \right],$$

where A is forest area, ω denotes the selective logging rate, and subscripts o, s, c, i, cv, f, and p stand for the area of original forest, selective logging, clear cutting, illegal logging, land conversion, forest fires, and plantation, respectively.

Finally, the forest-product volume H derived from selective logging and clear cutting is expressed as follows:

$$H_t = V_{s,t} + V_{c,t}.$$

Data

We obtained data relevant to the original forest area, forest-product volume, forest fire area, plantation area, and original forest density from the Ministry of Forestry (2008) (see Table 8.1). As for illegal logging volume and the land conversion area, the estimated figures were acquired from Chikyu-Ningen-Kankyo-Forum (2008: 7) and Tjondronegoro (2003), respectively. With these data as well as relevant calculated scores, the clear-cutting area is found based on the gap between the annual loss of forest area; subsequently, the selective logging area is also calculated on the basis of a selective logging rate $\omega = 50$ percent through reference to Chikyu-Ningen-Kankyo-Forum (2008: 101). We assume that the current forest density is close to the maximum forest density and therefore we set $D_{\max} = 60$ herein because the remaining forest is presumed to be in an almost climactic state due to more frequent clear cutting than selective logging. In terms of forest density in plantation areas, one cubic meter per hectare is applied, as extremely small figures are normally used in such a scenario.

Economic assessment model

Outline of the model

The economic assessment model in this study is a single-country, dynamic, CGE model for Indonesia. The economy consists of households, companies, and the government. The model employs the small country assumption for international

Table 8.1 Forest data in Indonesia

Item	Variable	Area (A)	Volume (V)	Density (D)
		1,000 ha	*1,000 m³*	*m³/ha*
Original forest	o	93,924[*3]	5,528,386[*1]	58.86[*2]
Annual forest loss	–	1,089[*3]	–	–
Selective logging	s	717	21,092[*1]	58.86[*2]
Clear cutting	c	179	10,558[*1]	58.86[*2]
Illegal logging	i	680[*1]	40,000[*4]	58.86[*2]
Land conversion	cv	226[*4]	13,295[*1]	58.86[*2]
Forest fire	f	5[*3]	278[*1]	58.86[*2]
Plantation	p	149[*3]	149[*1]	1.00[*2]

Notes:
[*1]: calculated from the density
[*2]: original figure
[*3]: Ministry of Forestry (2008)
[*4]: Chikyu-Ningen-Kankyo-Forum (2008).

trade; that is, the world prices of commodities are exogenously given. This model is dynamic, because households dynamically optimize their utility based on their future expectations of income, commodity prices, and factor prices. Company behavior is modeled as a conventional, static, profit-maximization problem for each time period. The model was developed using GAMS (General Algebraic Modeling System) software which has been widely used in CGE modeling. The model is formulated as a Mixed Complementarity Problem (MCP) and solved using the PATH solver. We conduct policy simulations from 2004 (the base year) to 2020 with one-year time steps.

Dynamic optimization of household problems

Households earn income by providing factors to production sectors, and they expend this income by either purchasing various commodities or by saving this income as household assets. This study distinguishes physical capital (material) and household assets (value, that is money); factor income from capital is treated differently from other factor incomes such as the wage income of labor. Other factor incomes are determined as a multiplication of supplied quantity (per capita) of factor f (denoted as x^f) and the price of factor f (denoted as w^f), while capital income is determined as a multiplication of per capita household assets (m), in value terms, and the rate of return (r). As a result, the per capita budget constraint of households is expressed as follows:

$$\sum_f w_t^f x_t^f + r_t m_t = \sum_f p_t^i c_t^i + S_t,$$

in which p^i is price of commodity i, c^i is per capita consumption of commodity i, and S is the quantity of household savings. The subscript t indicates the values in time t.

It is assumed that households produce a flow of satisfaction by consuming commodities as follows (Becker 1965):

$$c_t = \prod_i (c_t^i)^{\varphi i},$$

in which c is the produced flow of satisfaction and φ_i is a share parameter of commodity i in this household production function.

This study assumes that household utility at time t is determined by the discounted sum of felicities, where felicity is defined as satisfaction derived from current activities (Kojima 2007). Assuming a CIES (constant inter-temporal elasticity of substitution) felicity function $u(c_t) \equiv (c_t)^{1-\sigma}/(1-\sigma)$ in which σ is the inter-temporal elasticity of substitution, the household problem at time t is expressed as the following utility maximization problem:

$$\max_{\{c_t^i\}} U_t = \sum_{s=t}^{\infty} \left(\frac{1+v}{1+\rho}\right) u(c_s),$$

$$\text{subject to } m_{t+1} = \frac{m_t + S_t}{1+v} \quad \text{and} \quad \sum_f w_t^f x_t^f + r_t m_t = \sum_f p_t^i c_t^i + S_t,$$

in which ρ is the rate of pure time preference and v is the rate of population growth.

The following optimal consumption level is obtained as an interior solution to this household problem, while assuming the simplest expectation-formation process in which households assume that exogenous variables will remain constant at their current levels:

$$c_t^* = \left[\prod_i \left(\frac{\varphi^i}{p_t^i}\right)^{\varphi^i}\right] \times \left[1 + r_t - (1+v)\left(\frac{1+r_t}{1+\rho}\right)^{1/\sigma}\right] \times \left[m_t + \frac{1}{r_t - v}\sum_f w_t^f x_t^f\right].$$

For the policy simulation, this equation is extended to accommodate various taxes as follows:

$$c_t^* = \left\{\prod_i \left[\frac{\varphi^i}{(1+ts_t^i)p_t^i}\right]^{\varphi^i}\right\}$$

$$\times \left\{1 + (1-td_t)(1-tk_t)r_t - (1+v)\left[\frac{1+(1-td_t)(1-tk_t)r_t}{1+\rho}\right]^{1/\sigma}\right\}$$

$$\times \left[m_t + \frac{1-td_t}{(1-td_t)(1-tk_t)r_t - v}\sum_f (1+tf_t^f)w_t^f x_t^f\right]$$

in which ts^i is the sales tax on commodity i, tf^f is the tax on income from non-capital factor f, td is the direct tax on household income, and tk is the interest tax.

The optimal consumption level of each commodity is given as

$$c_t^{i*} = c_t^* \times \frac{\varphi^i}{(1 + ts_t^i)p_t^i} \prod_k \left[\frac{(1 + ts_t^k)p_t^k}{\varphi^k} \right]^{\varphi^k}.$$

Finally, the optimal household saving, S_t, is determined as the difference between household disposable income and expenditure for the optimal levels of consumption.

Companies

It is assumed that companies maximize profit through the production of goods and services by inputting intermediate commodities as well as production factors, given the production technology. The production technology is assumed to be described by a Leontief function of intermediate goods, and a CES (constant elasticity of substitution) function of factors of production (labor and capital) which reflects imperfect substitutability between factors. The model assumes a perfect market in which a market equilibrium solution means that demands and supplies of all commodities and factors are balanced and company profits become zero.

Production sectors are aggregated into six sectors focusing on forestry-related sectors: forestry (frs), wood products (lum), agriculture and fishery (xag), mining (xmn), other manufacturing (xmf), and services (xsv). Production factors are skilled labor, unskilled labor, capital, and land and natural resources, among which labor and capital are assumed to be mobile across sectors, while land and natural resources are sector-specific. It is assumed, as mentioned above, that the natural resource input to the forestry sector corresponds to logging volume in the forest-stock model. The endowments of other natural resources and land are fixed at their base year levels. The growth rates of skilled and unskilled labor endowment are set according to macroeconomic projections by the Center for Global Trade Analysis at Purdue University. Capital accumulation in this model is endogenously determined, based on an assumption that all household saving is invested in capital accumulation.

Produced goods and services are allocated to either the domestic market or exports according to a CET (constant elasticity of transformation) function. Further, we employ an assumption of imperfect substitution between domestically produced commodities and imported commodities in the same category (Armington 1969).

Government

The government not only represents the tax transactions and consumption and investment of the government recorded in the base-year data in the model, but it also receives REDD credits as well as PES and dispends for afforestation under the SFU scenarios. The model assumes that the government maintains budget neutrality through lump-sum transfers of the budget surplus to households. Consumption and investment by the government are not reflected in the utility function, and thus are assumed to be constant at their base-year levels.

Equilibrium conditions and macro closure rules

The markets for commodities and factors are assumed to be cleared through the endogenous determination of equilibrium prices. As mentioned before, we distinguish physical capital, which is material subject to wear and tear, from households' assets in terms of monetary value; the factor price of capital does not coincide with rate of return to households' assets. The relationship between households' return to assets and capital costs for companies can be described as follows:

$$N_t r_t m_t = \sum_j w_t^K z_t^{jK} - p_t^K \delta k_t.$$

Here, N_t is population, w_t^K is the factor price of capital, z_t^{jK} is the capital input of company j, δ is the depreciation rate, p_t^K is the price of capital goods, and k_t is the capital stock. Per capita household assets (m) and the capital stock of the whole economy (k) are described by the following equations, respectively:

$$m_{t+1} = \frac{(m_t + S_t)}{1 + v},$$

$$k_{t+1} = (1 - \delta)k_t + I_t,$$

where I_t is investment in capital which satisfies $N_t S_t = p_t^K I_t$, as we assume equality between total saving and capital investment in each year.

These relationships, as well as the market-clearance conditions for capital, determine the factor price of capital and the rate of return endogenously.

The budget neutrality of the government is achieved through endogenous adjustment of lump-sum transfers from the government to households, given public investment, except for afforestation and government consumption which are exogenously fixed at their base-year levels. The closure rule concerning the balance of the current account and the exchange rate is endogenously determined, such that the balance of the current account is fixed at its base-year level.

Data

We constructed a social accounting matrix (SAM) for Indonesia based on the GTAP database version 7, the widely used world economic and trade data for 2004. This study aggregates the GTAP database version 7 – which consists of 57 sectors and 108 regions – into two regions (Indonesia and the rest of the world) and the six sectors mentioned above for dealing with Indonesian forestry-related sectors.

The GTAP database deals with the lack of data for some transactions related to households and the government by introducing a hypothetical "regional household" account. This collects all factor incomes and tax revenues, saves a part of revenues, and transfers the remaining revenues to households and the government. This specification makes it difficult to conduct policy simulations assuming the government's budget neutrality, for example a simulation involving public expenditure from tax revenues while keeping budget neutrality. Here, we solve this problem by eliminating the regional household account from the SAM as follows.

Table 8.2 shows transactions related to the regional household account in the SAM including the regional household account. In this table, the rows indicate revenues and the columns indicate expenditures. For example, Section A indicates direct taxes paid by production factors, and Section F indicates a transfer from the regional household to households. It is not arguable that Section B, the factor income of the regional household, is in reality the factor income of households, and that the tax revenues of the regional household (Sections D and E) are those of the government. The remaining information necessary to eliminate the regional household account is the direct tax paid by households (x_1) and the distribution of regional household savings (H) to household savings (x_2) and government savings (x_3).

These unknown variables can be determined by setting the share of private savings (σ) equal to the total savings as shown in Table 8.3.

Policy scenarios

We formulate policy scenarios starting from 2004, the base year of the economic assessment model. In order to reflect the negative impacts of forest loss on economic activities, we assume that the unit logging effort would be increased in inverse proportion to forest decrease. In more detail, effort is set to rise by 25 percent and 50 percent when forest areas have diminished by 30 percent and 50 percent, respectively (subsequently approximated by an exponential function). Furthermore, as selective logging is considered to require more effort, the unit

Table 8.2 Transactions related to the regional household account in the SAM

	FF	ITAX	DTAX	REG	HH	GOV	SAV
Production factors (FF)							
Indirect tax (ITAX)					I	J	K
Direct tax (DTAX)	A						
Regional households (REG)	B	D	E				
Households (HH)				F			
Government (GOV)				G			
Savings (SAV)	C			H			

Table 8.3 Determined transactions without the regional household

	FF	ITAX	DTAX	HH	GOV	SAV
Production factors (FF)						
Indirect tax (ITAX)				I	J	K
Direct tax (DTAX)	A			$B - F - \gamma H$		
Households (HH)	B					
Government (GOV)		D	$E + \{B - F - \gamma H\}$			
Savings (SAV)	C			γH	$(1 - \gamma)H$	

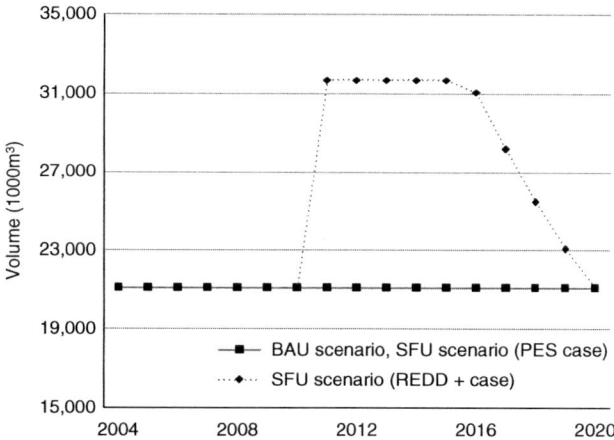

Figure 8.1 Annual selective logging volume in each scenario.

logging effort thereof is set at 1.44 times higher than that of clear cutting. This is estimated from the difference between the costs of conventional and low-impact logging (referring to Dagang *et al.* 2002).

BAU scenario

The BAU scenario reflecting the current trend of forest loss in Indonesia presumes that existing selective logging, clear cutting, illegal logging, land conversion, forest fires, and plantation would be continued until 2020 (see Figures 8.1 to 8.4). Figure 8.5 indicates the trend of selective logging areas and Figure 8.6 shows the trend of plantation area, assuming a 30 percent survival rate over the three-year period. The total forest area at the end of each year is illustrated in Figure 8.7; the total forest stock is presented in Figure 8.8. Existing logging activities keep annual forest production constant until 2020 (see Figure 8.9), while annual CO_2 emissions would gradually decrease owing to forest growth in the area of the original forests, selective logging, and plantation (see Figure 8.10).

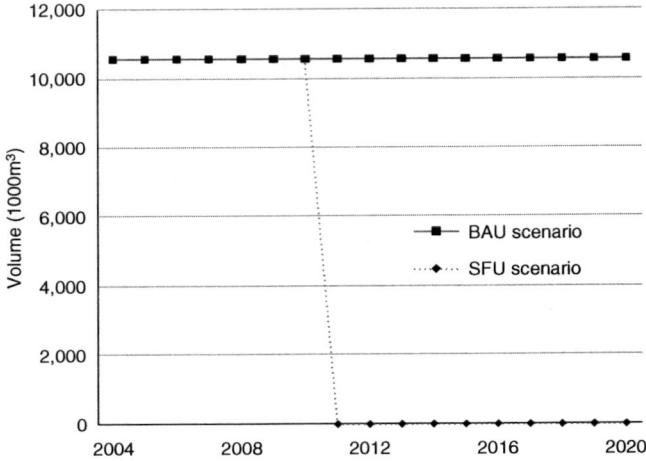

Figure 8.2 Annual clear cutting volume in each scenario.

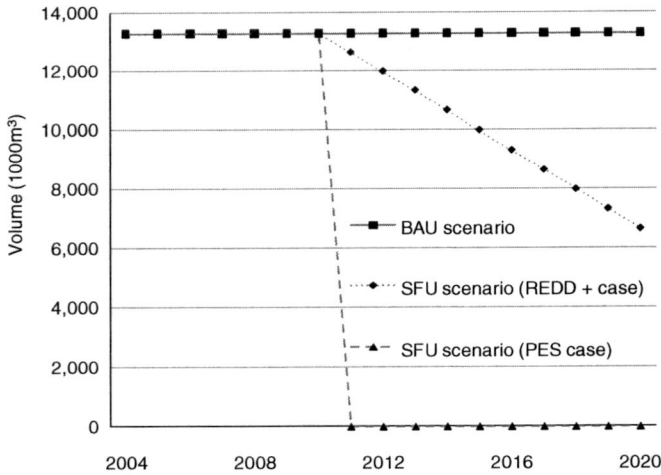

Figure 8.3 Annual forest volume loss due to land conversion in each scenario.

Sustainable forest-use (SFU) scenarios

The SFU scenario assumes that all clear cutting would be transformed into selective logging, and land-conversion areas would decrease linearly by 50 percent from their year 2010 levels by 2020 (see Figures 8.1 to 8.3). Also, plantation areas are presumed to increase linearly to cover the loss of forest areas due to illegal logging and forest fires in the status quo, setting a 40 percent survival rate in three-year time periods (referring to BAPPENAS 2009) (see Figure 8.4).

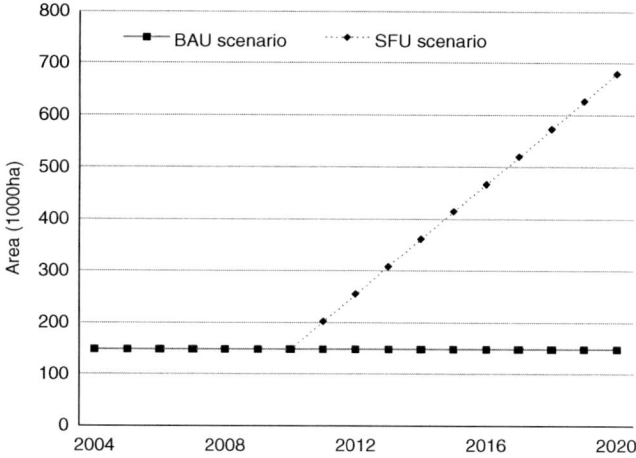

Figure 8.4 Annual plantation area in each scenario.

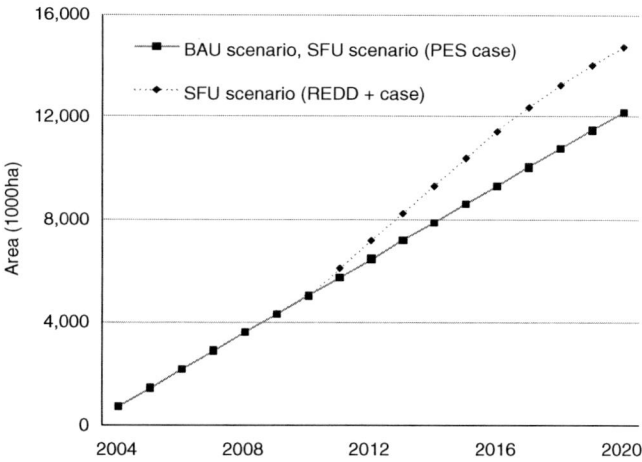

Figure 8.5 Total selectively logged area in each scenario.

Figures 8.5 to 8.8 illustrate the trend of selective logging area, plantation area, total forest area, and total forest stock in this scenario. Due to transformation of logging practices, the annual forest production will be less than that in the BAU scenario (see Figure 8.9).

The SFU scenario will achieve the CO_2 emissions-reduction target in Indonesia. Based on the target in the forestry sector (Ministry of Environment 2010), this scenario assumes that the CO_2 emissions will be reduced gradually from

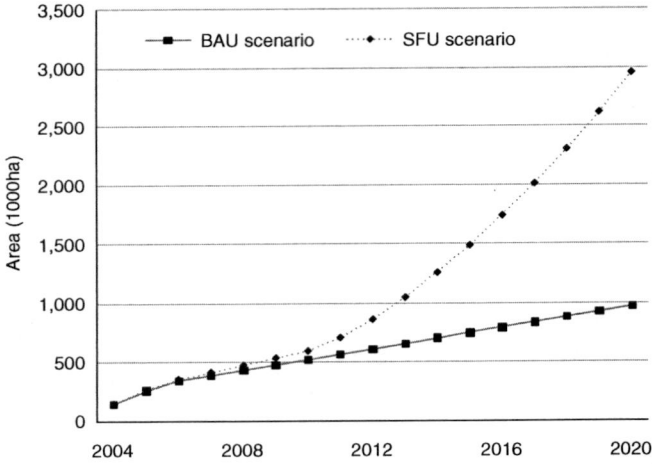

Figure 8.6 Total plantation area in each scenario.

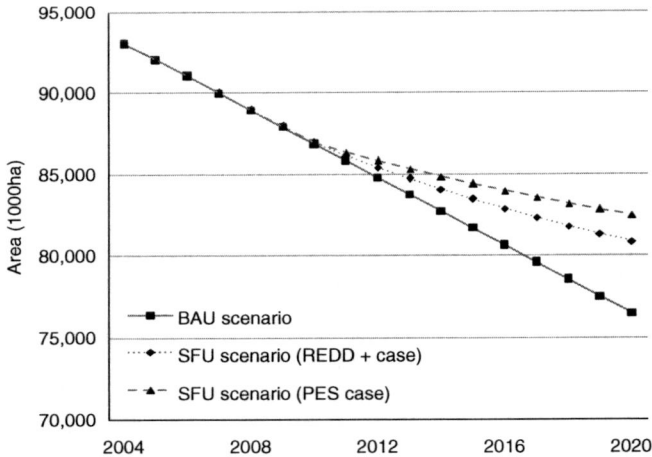

Figure 8.7 Total forest area at each end of year in each scenario.

2011 by 42.8 percent of the BAU scenario by 2020 (see Figure 8.10). Although logging itself does not necessarily emit CO_2 immediately, all the logged wood volumes will be converted into CO_2 emissions in this scenario, in line with the discussion in the United Nations Framework Convention on Climate Change (UNFCCC). Nevertheless, the emissions from selective logging are considered

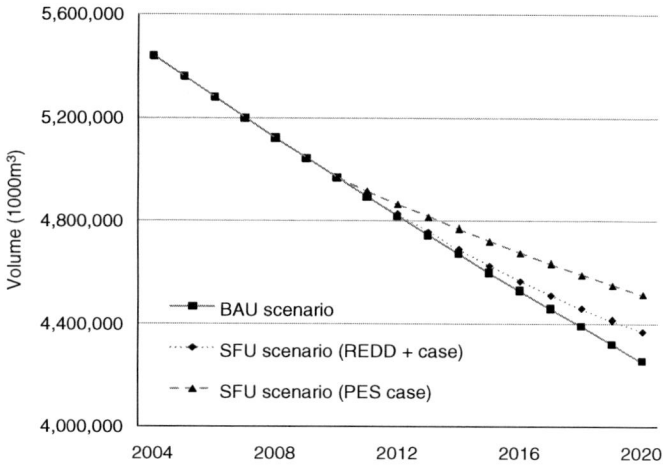

Figure 8.8 Total forest stock at each end of year in each scenario.

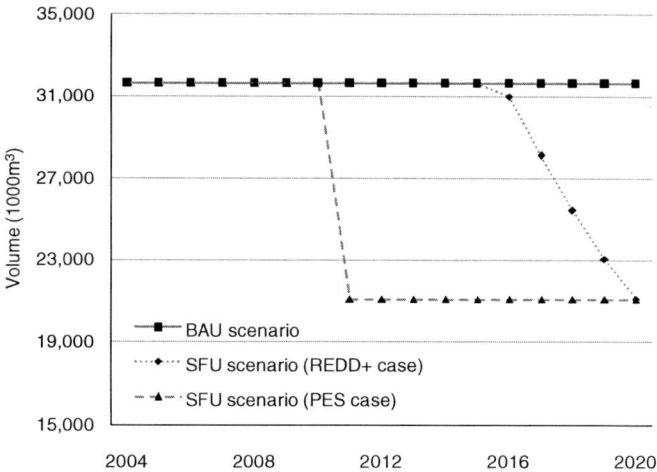

Figure 8.9 Annual forest production in each scenario.

to be smaller than those from clear cutting, thereby adjusting those from selective logging to be smaller, by multiplying the selective logging rate. This is because remaining woods are supposed to keep storing carbon in the soil, while clear cutting will release carbon from the soil. As a result of this scenario which reduces selective logging volumes and increases plantation areas for fulfilling the

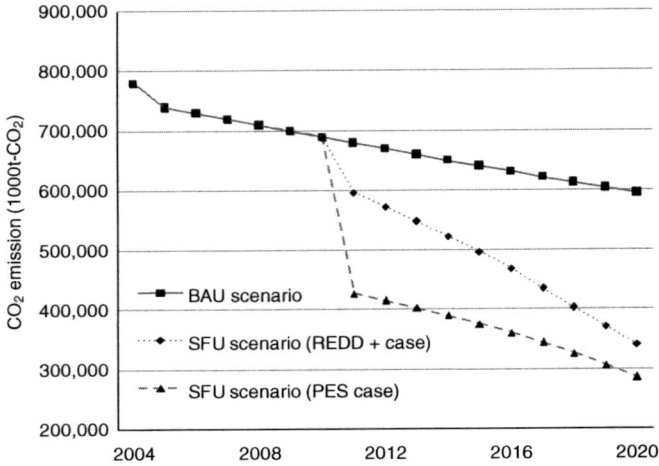

Figure 8.10 Annual CO_2 emission in each scenario.

emission-reduction target, the income in the forestry sector will decrease while the cost is set to increase.

Meanwhile, REDD credits will be issued in this scenario, which are equivalent to CO_2 emission reduction, in order to prevented forest loss compared to the BAU scenario. Keith *et al.* (2009) estimate the carbon volume contained in one hectare of tropical forest at 171 t-C, implying 627 t-CO_2 per unit hectare from the conversion equation between carbon and CO_2, and 10.45 t-CO_2 per unit wood volume when applying $D_{max} = 60$. From the above and the credit price of 4 US dollars per t-CO_2 applied in the IFCA (2008), the credit equivalent to 41.8 US dollars is issued to one cubic meter from the year 2011 and beyond, when the difference in CO_2 emission from the BAU scenario can be observed. The income from REDD credits is transferred to domestic households herein to keep the financial balance of the government constant.

Simulation results

Social welfare

We assess the social welfare impacts of scenarios in terms of equivalent variation (EV) that is defined as the amount of compensation for giving up the assessed policy without reducing utility levels. Figure 8.11 shows EV in each time step during the simulation period.

Throughout the simulation period, the SFU scenario is associated with negative EV because negative economic impacts due to the reduction in forestry production exceed the positive impacts due to the REDD credit revenues. The

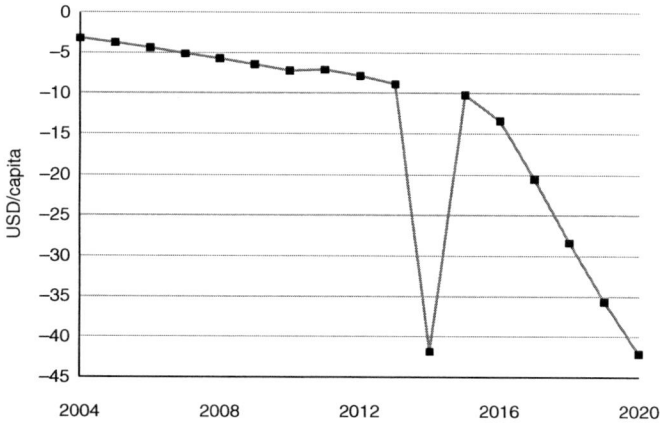

Figure 8.11 Equivalent variation (EV) of SFU.

net present value of EV for the entire simulation period is a negative 110 US dollars per capita. The simulation results indicate that SFU reduces the social welfare level. It must be noted that this study does not reflect the potential benefits of the SFU scenario in achieving the CO_2 emission-reduction target, which is expected to improve social welfare. Similarly, the potential social welfare improvement of enhanced sustainability of forest ecosystem and forestry due to SFU is not taken into account.

Economic impacts

Figure 8.12 shows the time path of real GDP under the BAU and the SFU scenarios.

It shows that SFU has negative impacts on GDP throughout the simulation period, as the negative impacts of production reduction in the forestry sector exceed the positive impacts of the revenues from the REDD credit. These results are further illustrated by Figures 8.13 and 8.14 which show a significant fall in production in the forestry and the wood products sectors.

While the production of the forestry sector, even under the BAU scenario, gradually reduces year by year due to the increased production costs caused by the reduced forest stock, the scale of reduction under the SFU scenario that restricts the logging volume is much larger. It is interesting that the scale of production reduction is much larger in the wood-products sector – which inputs forestry products as a major intermediate input – than it is in the forestry sector that is directly affected by SFU (according to the comparison of Figures 8.13 and 8.14). Figure 8.15 shows the impacts of SFU (percentage change from the BAU) on forestry-product intermediate inputs to each sector.

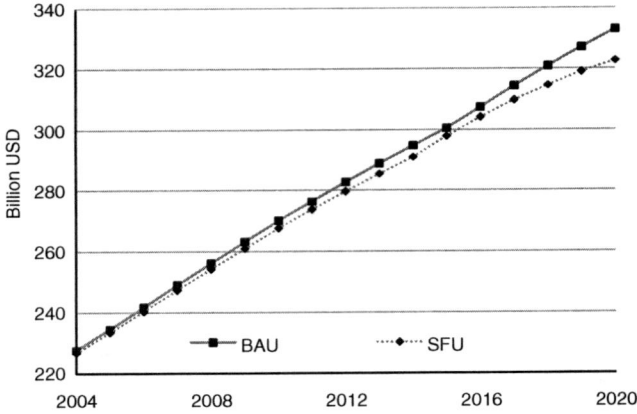

Figure 8.12 Real GDP.

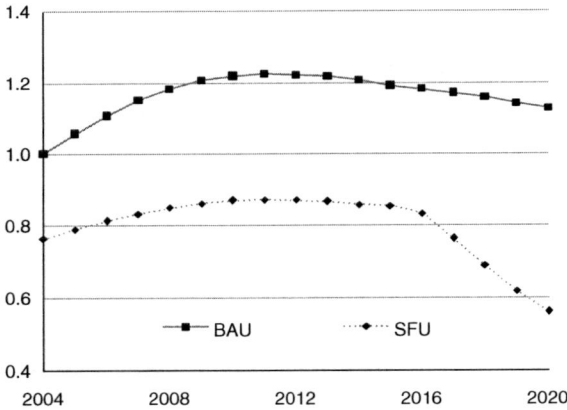

Figure 8.13 Forestry sector production (BAU in the base year = 1).

While the forestry-product intermediate inputs to the wood-product sector (lum) fall drastically from −40 to −65 percent, those to other sectors do not fall significantly. It is found that the impacts of the reduced supply of forestry products due to SFU are concentrated in the wood-product sector that inputs these products as its major intermediate input.

Impacts on households

Figure 8.16 shows the time path of households' income under the BAU and the SFU scenarios. Even though the REDD credit revenues are finally transferred to

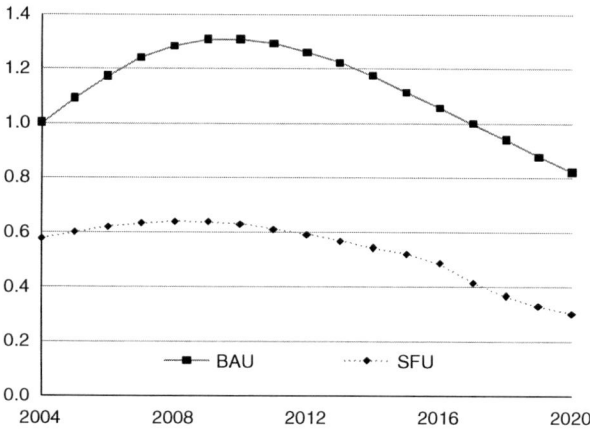

Figure 8.14 Wood products sector production (BAU in the base year = 1).

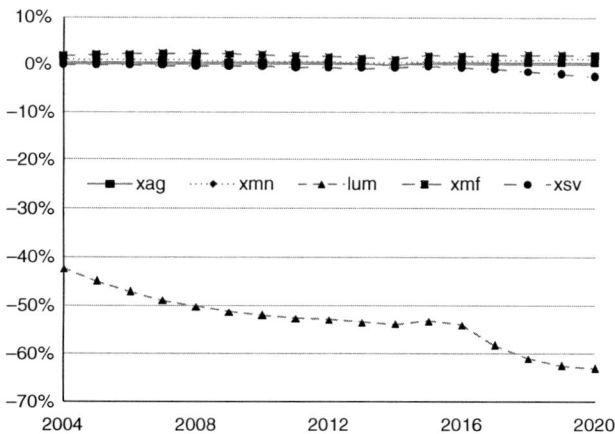

Figure 8.15 Impacts of SFU on intermediate input of forestry products.

households, SFU reduces households' income throughout the simulation period, except for 2014.

A unique feature of this model is an endogenous determination of household savings according to dynamic optimization. Figure 8.17 shows the time path of households' propensity to save, that is, the ratio of savings to total income.

SFU reduces the propensity to save throughout the simulation period, except for 2014. As a result, SFU reduces the households' assets as shown in Figure 8.18.

The households' assets are a source of future utility, and their accumulated quantity at the end of the simulation period is an important assessment indicator,

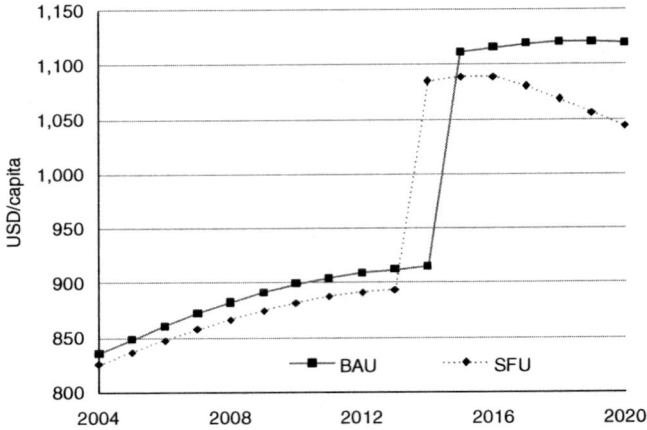

Figure 8.16 Household income.

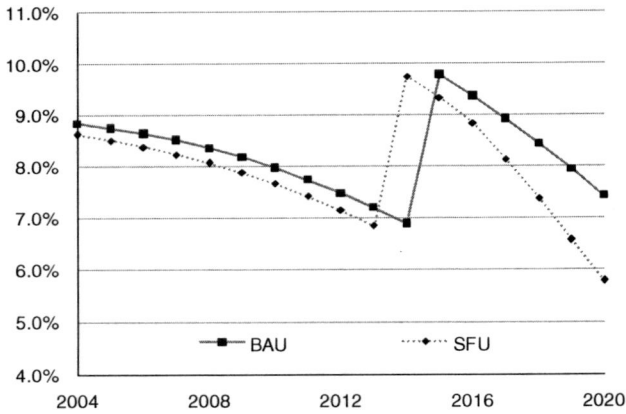

Figure 8.17 Household propensity to save.

as well as showing social welfare levels during the simulation period. The simulations of this study predict reduced social welfare levels during the simulation period and a reduction in households' assets at the end of the simulation period by 1.3 percent from the BAU level, which are unfavorable to SFU.

REDD credit price and PES price for maintaining the social welfare level

The both REDD and PES have attracted attention as innovative financing mechanisms to promote sustainable ecosystem service use, but there are some remaining

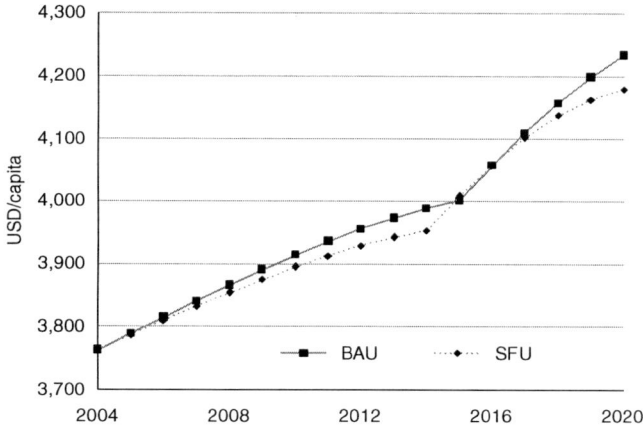

Figure 8.18 Households' assets.

problems to be solved for their wide application. In particular, appropriate pricing of ecosystem services is very important since these financial mechanisms generate money by assigning a monetary value to these services which are traditionally not transacted in the market. Even though the mainstream approach to tackle this issue is to estimate benefits of ecosystem services based on environmental-valuation techniques, this study takes an alternative approach to estimate the REDD credit price (in the REDD case) or the PES price for forest ecosystem services, other than the carbon-fixation function (in the PES case), such that the SFU scenario does not lower social welfare levels.

In the REDD case, we seek the REDD credit price for which the net present value of EV of the SFU scenario is not negative. While the base case (BL) assumes the REDD credit price at 4 US dollars per t-CO_2, increasing the REDD credit price by around nine times (35.6 US dollars per t-CO_2) results in a non-negative net present value of EV (8.7 US dollars per capita). Please note that this model is strongly nonlinear and the price slightly lower than that price results in negative net present value of EV.

In the PES case, we seek the PES price for forest ecosystem services except for the carbon-fixing function with which the net present value of EV of the SFU scenario is not negative. Here we assume that the REDD credit price is the same as the BL case (4 US dollars per t-CO_2). In the PES case, all clear-cut logging and land-development activities are halted in exchange for PES with a fixed amount per unit area of halted activities. The volume of selective-cut logging is fixed at the base-year level and the afforestation area is assumed to increase linearly until 2020, assuming a 40 percent three-year survival rate. Under these assumptions, the net present value of EV becomes non-negative, 0.9 US dollars per capita, when the PES price is 8,500 US dollars per hectare. The total PES revenue with this price is around 3.59 billion US dollars per year for the simulation period. For reference, in the case of Costa Rica, one of the most advanced countries in PES, the users of ecosystem services paid 57 million US dollars as PES to the land owners for the five years between 1997 and 2001 (Malavasi and Kellenberg 2002). Indonesia is 36 times larger in area and 56 times larger in population than is Costa Rica.

Figure 8.19 shows the time path of EV of SFU in each case during the simulation period.

There is no significant difference in the time path of EV between the REDD case and the PES case, in spite of their difference in financial inflow from foreign countries.

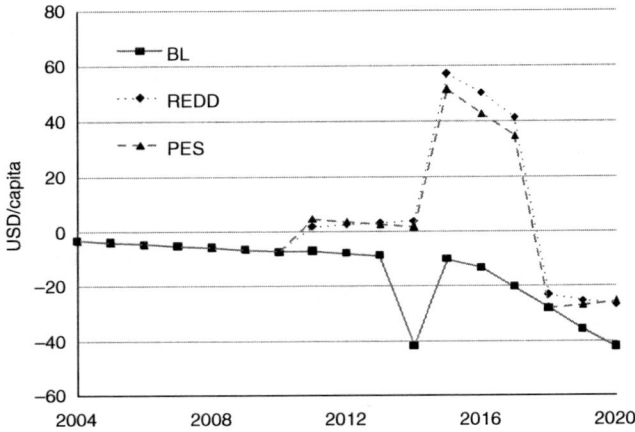

Figure 8.19 EV of SFU in each case.

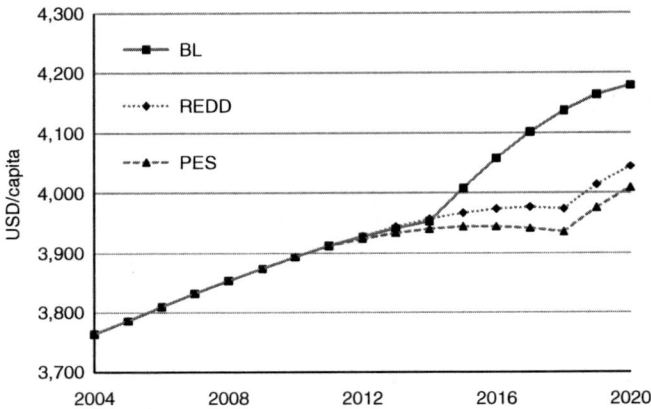

Figure 8.20 Households' assets in each case.

As shown in Figure 8.20, SFU results in a reduction in the households' assets at the end of the simulation period, which are the source of generating future utility after the simulation period, in both the REDD case (−4.5 percent) and the PES case (−5.3 percent).

According to these results, SFU reduces the households' assets (man-made capital) but conserves forest stock (natural capital) under the condition of maintaining the social welfare level. This study implies potential tradeoffs between natural and man-made capital in the SFU policy.

Conclusions

This chapter presents an attempt to assess the policy impacts of sustainable ecosystem service use quantitatively. We develop an assessment tool combining a forest-stock model, which reflects afforestation, selective and clear-cut logging, and natural growth of forest, and a multisectoral Ramsey-type growth model (a dynamic CGE model), and we apply these tools to assess the policy impacts of SFU policies in Indonesia quantitatively.

Unlike the existing studies that employ CGE models to assess the REDD from the view of impacts on the carbon-credit market or conduct scenario analysis in order to assess environmental impacts of SFU using dynamic input–output models (cf. Hamilton 1997), this study assesses the economic impacts of the SFU scenarios, which are specified based on a forest-stock model, using a dynamic CGE model which treats capital (and households' assets) as an endogenous variable. In our scenarios, SFU is defined as a restriction of logging volume such that the forest stock is maintained and the SFU scenarios assume the employment of reduced-impact logging (RIL) technologies which are associated with higher production costs than are the conventional ones. This study considers negative impacts of forest-stock depletion through increased efforts per unit volume of logging. The REDD and PES are regarded as financing mechanisms of SFU policies, and the revenues from these schemes are either spent for afforestation or are transferred to households.

In the base case, setting the REDD credit price at 4 US dollars per $t\text{-}CO_2$ according to the existing studies, the SFU policy reduces the social welfare level in terms of EV. This negative impact is mainly caused by the production reduction in the forestry sector due to restrictions in logging volume, and the wood-product sector that inputs the forestry products as the main intermediate input. In this analysis, the capital stock and its underlying households' assets are endogenously determined, and the SFU scenario results in 1.3 percent less households' assets at the end of the simulation period than the BAU scenario does. This means that the SFU can conserve natural capital (forest stock) but reduce man-made capital and its underlying households' assets.

This study also deals with the issue of the appropriate pricing of the REDD credit and PES. The REDD credit price or the PES price are endogenously determined, such that the SFU scenario does not reduce social welfare levels. Our results indicate that such a REDD credit price is 35.6 US dollars per $t\text{-}CO_2$, around nine times higher than the current 4 US dollars per t-CO_2, and that such a PES price for forest ecosystem services other than a carbon-fixation service is 8,500 US dollars per hectare, which is equivalent to 3.59 billion US dollars per year of total PES revenues. In either case, the SFU scenario results in less households' assets at the end of simulation period than the BAU scenario, 4.5 percent less in the REDD case, and 5.3 percent less in the PES case.

These simulation results highlight an important point for conducting a policy-impact assessment of sustainable ecosystem service use including SFU, that is, how to reflect the impacts of policy on the terminal stock of various types of capital including natural capital to the assessment, in addition to those on the social welfare levels during the simulation period. This is particularly important when there is a tradeoff between different types of capital.

Last, we would like to mention some of the remaining tasks. The utility function employed in this study does not include the benefits of ecosystem services and we employ the assumption that REDD and PES are paid by foreign countries in order to reflect these revenues directly affecting social welfare levels. The introduction of ecosystem service benefits into the utility function will enable us to capture the social welfare benefits of domestic transactions of PES. Elaboration of the forest stock model is another important issue. Employment of the Faustmann model that takes account of the rotation of afforestation and logging will contribute to elaborating assumptions behind the policy scenarios; adding more SFU policy options using the revenues from REDD or PES will make it possible to conduct policy-impact assessment more relevant to actual policy needs.

References

Ahammad, H. and Mi, R., 2005, *Land Use Change Modeling in GTEM: Accounting for Forest Sinks*, ABARE Conference Paper, No. 05.13.

Armington, P. S., 1969, "A theory of demand for products distinguished by place of production," *IMF Staff Papers*, 16: 159–176.

BAPPENAS, 2009, *The Indonesia Climate Change Sectoral Roadmap: Synthesis Report*, BAPPENAS, Republic of Indonesia.

Becker, G. S., 1965, "A theory of the allocation of time," *Economic Journal*, 75(299): 493–517.

Chikyu-Ningen-Kankyo-Forum, 2008, *Ihobassai ni yoru kankyou eikyou chousa gyoumu* (Research Project on Environmental Impact of Illegal Logging), report commissioned from Ministry of Environment Japan, Chikyu- Ningen-Kankyo-Forum (in Japanese).

Chopra, K. and Kumar, P., 2004, "Forest biodiversity and timber extraction: An analysis of the interaction of market and non-market mechanisms," *Ecological Economics*, 49(2): 135–148.

Dagang, A. A., Richter, F., Hahn-Schilling, B. and Manggil, P., 2001, *Financial and Economic Analyses of Conventional and Reduced Impact Harvesting Systems in Sarawak*, paper read at the International Conference Proceedings on Applying Reduced Impact Logging to Advance Sustainable Forest Management, 26 February to 1 March 2001, Kuching Sarawak, Malaysia.

Engel, S. and Palmer, C., 2008, "Payments for environmental services as an alternative to logging under weak property rights: The case of Indonesia," *Ecological Economics*, 65(4): 799–809.

Forest Watch Indonesia and Global Forest Watch, 2002, *The State of the Forest: Indonesia*, Washington, DC: Global Forest Watch.

Goodland, R. J. A., Asibey, E. O. A., Post, J. C. and Dyson, M. B., 1991, "Tropical moist forest management: The urgency of transition to sustainability," in I. R. Costanza (ed.) *Ecological Economics: The Science and Management of Sustainability*, New York: Columbia University Press, pp. 486–516.

Hamilton, C., 1997, "The sustainability of logging in Indonesia's tropical forests: A dynamic input–output analysis," *Ecological Economics*, 21(3): 183–195.

Indonesia Forest Climate Alliance (IFCA), 2008, *Reducing Emissions From Deforestation and Forest Degradation in Indonesia: REDD Methodology and Strategies Summary for Policy Makers*, IFCA, Ministry of Forestry, Republic of Indonesia.

Keith, H., Mackey, B. G. and Lindenmayer, D. B., 2009, "Re-evaluation of forest biomass carbon stocks and lessons from the world's most carbon-dense forests," *PNAS* 106(28): 11635–11640.

Kojima, S., 2007, *Sustainable Development in Water-stressed Developing Countries: A Quantitative Policy Analysis*, Cheltenham: Edward Elgar.

Malavasi, E. O. and Kellenberg, J., 2002, *Program of Payments for Ecological Services in Costa Rica*, presented at Building Assets for People and Nature: International Expert Meeting on Forest Landscape Restoration, Heredia, Costa Rica, 27 February to 2 March 2002.

Millennium Ecosystem Assessment, 2005, *Ecosystems and Human Well-Being: Synthesis*, Washington, DC: Island Press.

Ministry of Environment, 2010, *NAMAs and MRV Manner*, Ministry of Environment, Republic of Indonesia, Presentation material at The 19th Asia-Pacific Seminar on Climate Change: Toward Low Carbon and Climate Changes-resilient Asia-Pacific-From Copenhagen to Mexico, 20–22 July, 2010 Kitakyushu, Japan.

Ministry of Forestry, 2008, *Forestry Statistics of Indonesia 2007*, Ministry of Forestry, Republic of Indonesia, available at http://www.dephut.go.id/index.php?q=en/node/6123 (Accessed: 6 February 2012).

Nordhaus, W. D., 1993, "Rolling the "DICE": An optimal transition path for controlling greenhouse gases," *Resource and Energy Economics*, 15: 27–50.

Sohngen, B. and Mendelsohn, R., 2003, "An optimal control model of forest carbon sequestration," *American Journal of Agricultural Economics*, 85(2): 448–457.

Tjondronegoro, S. M. P., 2003, *Land Policies in Indonesia*, EASRD Working Paper for a Land Policy Dialogue, Jakarta.

Washida, T., 1999, "Simulation analysis including ecosystem assessment using a global environmental economic model," *Kokumin Keizai Zasshi* (Journal of National Economy), 179(5): 43–60 (in Japanese).

World Commission on Environment and Development, 1987, *Our Common Future*, Oxford: Oxford University Press.

9 Financing REDD-plus

A review of options and challenges

Kimihiko Hyakumura and Henry Scheyvens

Introduction

In this chapter we provide an overview of the options available and the challenges for the financing of REDD-plus readiness and its implementation. REDD refers to reducing emissions from deforestation and forest degradation, and the plus refers to the role of conservation, sustainable management of forests, and enhancement of forest carbon stocks in developing countries. REDD-plus is now a hot topic on the international climate change agenda, and investment for readiness and demonstration activities are taking place in over 30 developing countries. This chapter provides an introduction to the REDD-plus concept, discusses the types of costs for REDD-plus as well as possible types of financing, and concludes with a number of observations regarding the complexities and challenges facing REDD-plus financing.

Background

Emissions of carbon dioxide (CO_2) from deforestation and forest degradation have attracted global attention in recent years. The Intergovernmental Panel on Climate Change (IPCC) Fourth Assessment Report (AR4) notes that CO_2 emissions due to changes in land use caused by deforestation and forest degradation amount to around 20 percent of total anthropogenic CO_2 emissions (IPPC 2007) (see Figure 9.1). This makes deforestation the second largest anthropogenic source of CO_2 emissions. The Stern Review (Stern 2006) concluded that implementing preventive measures for deforestation and forest degradation, compared with other preventive measures for global warming, is "a highly cost-effective way of reducing greenhouse gas emissions." Attention must focus on some developing tropical countries, such as Brazil, Indonesia, and Nigeria, where deforestation rates are highest, as shown in Figure 9.2 (FAO 2010).

With this backdrop, Parties to the United Nations Framework Convention on Climate Change (UNFCCC) are developing a new mitigation mechanism for global warming known as REDD-plus that aims to provide incentives to developing countries to protect and enhance their forest carbon stocks (see Figure 9.3). REDD-plus became a hot topic in the international climate change negotiations after Papua New Guinea and Costa Rica presented their concept of "Avoided

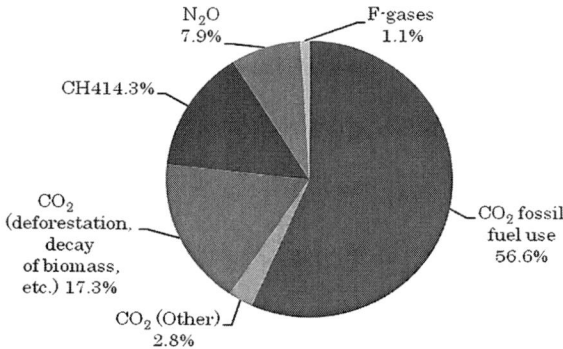

Figure 9.1 Global anthropogenic GHG emissions in 2004.

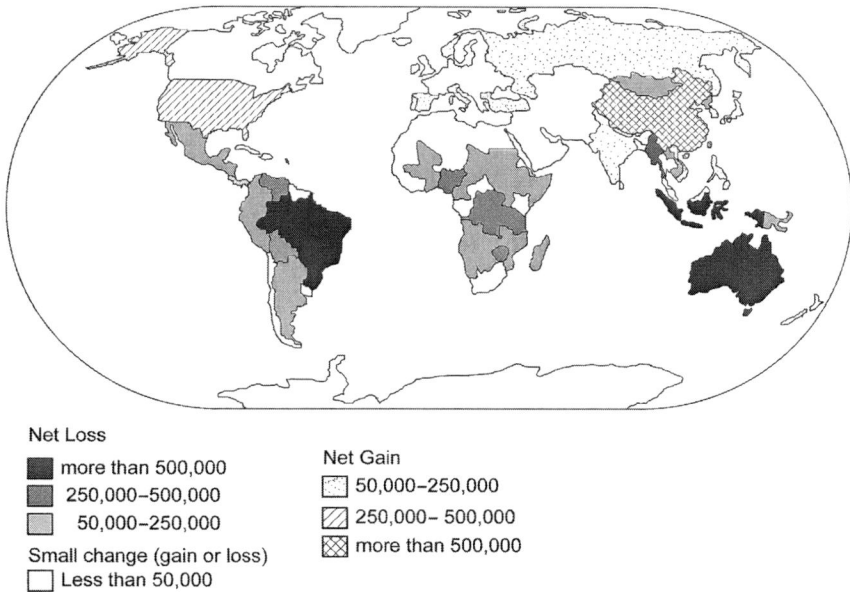

Net Loss
■ more than 500,000
▨ 250,000–500,000
▨ 50,000–250,000

Net Gain
▦ 50,000–250,000
▨ 250,000– 500,000
▨ more than 500,000

Small change (gain or loss)
☐ Less than 50,000

Figure 9.2 Countries with large net changes in forest.

Deforestation" at the eleventh session of the Conference of the Parties (COP) to the UNFCCC in 2005. By the thirteenth COP in 2007, REDD-plus had become one of the main agenda items. The Bali Action Plan, an outcome of this COP, stated that REDD-plus should be considered in the future global-climate framework, and expanded the concept of REDD to include the plus activities described in the introduction above. Taking their lead from the thirteenth COP, and with the support of

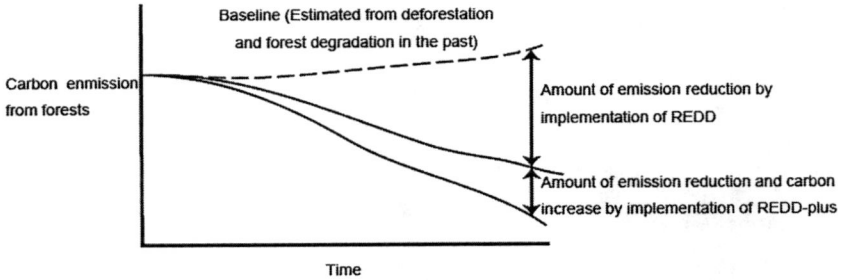

Figure 9.3 REDD-plus outline.

international organizations, developing countries began preparing themselves for REDD-plus. REDD-plus was further elaborated by the COP at is sixteenth and seventeenth sessions. In 2011, the seventeenth COP confirmed the REDD-plus activities and agreed that REDD-plus should be implemented in phases, moving from preparation to actions that are fully measured, reported, and verified, and with a set of seven social and environmental safeguards. While the negotiations are still to deal with several difficult issues, such as how performance-based REDD-plus activities should be financed, and the relationship between REDD-plus and agriculture, preparations at national and sub-national levels, as well as demonstration activities, are well underway.

One factor behind the strong interest in REDD-plus is speculation about possible gains by both developed and developing countries. Developing countries are hoping to receive new investment for forest management from multilateral and bilateral sources. They expect that REDD-plus will provide co-benefits, such as proper national forest inventories, which are yet to be conducted in some countries, and better forest monitoring. Developed countries, on the other hand, are expecting to reduce the costs of achieving mitigation targets by having access to a new cheap source of emissions offsets. In spite of difficult methodological and political issues that still need to be resolved and agreed, the strong interest from developing and developed countries, and the fact that REDD-plus is already relatively well-advanced in their negotiations, suggest that REDD-plus will be realized as part of the future global climate framework.

While the concept of REDD-plus appears straightforward, implementation will be complex and challenging. While REDD-plus focuses on carbon, the reduction of forest management to the management of carbon stocks poses serious risks. Forests provide multiple functions, not just carbon sequestration and storage. They contribute to watershed and soil protection, biodiversity conservation, and economic development, by providing timber and non-timber forest products, and they provide resources important for the subsistence of local people. There is much hope that REDD-plus will secure these various functions (Brown *et al.* 2008), especially as in recent years global funding for forest conservation has been declining (Khare *et al.* 2005). However, a narrow approach to REDD-plus could be to the detriment of some of the functions, which explains why the COP agreed to a set of safeguards for REDD-plus. For example, to protect the rights and well-being of local people who depend upon forests and forest lands, a safeguard that requires the participation of local communities and indigenous people in all phases of REDD-plus was introduced. Transparent governance is another of the safeguards, but this, too, is not so straightforward. Governments have been devolving rights to forest management over the past several decades, but REDD-plus could act against this devolution if central governmental authorities seek to capture the benefits from carbon sales (Phelps *et al.* 2010). This complexity, combined with the potentially large international resource flows that could be generated, explains the strong interest in REDD-plus, not only from those involved directly in REDD-plus preparations and negotiations, but also from non-forestry government sectors, aid agencies, non-governmental organizations (NGOs), and researchers.

Types of REDD-plus costs

Negotiators have agreed that REDD-plus results-based actions should be fully measured, reported, and verified. They have also agreed that developing countries should devise systems for reporting information on the implementation of the safeguards. REDD-plus thus demands new capacities beyond those required for conventional forest management or conservation. Investment in remote sensing technologies and ground-based measurement and carbon-stock estimation can be seen in the countries now preparing for REDD-plus. Readiness also requires investment in organizations to manage the readiness process and to implement REDD-plus, national reviews of deforestation drivers and policy performance, awareness raising, multi-stakeholder dialogues, and new policies, regulations, and standards. To these must be added costs associated with implementing and evaluating the REDD-plus activities and compensation costs (also referred to as positive incentives) for implementing the activities (see Table 9.1).

Looking more closely at readiness costs, we see that as forest carbon stock must be measured and monitored there will be costs for developing and implementing remote-sensing and ground-based forest monitoring methods. Costs will also be associated with the process of developing appropriate mechanisms to share the benefits from the results-based actions. There is another set of costs associated with the development of national REDD-plus strategies or road maps, which the COP has requested developing countries to prepare. In addition, costs will be incurred for building the capacity of forestry departments and others involved in REDD-plus design and implementation.

How much will this all cost? We have some indicative figures. About 50 million US dollars is required just for making the forest-resources inventory in 25 developing countries (Eliasch 2008). Capacity building for REDD-plus in 40 countries could cost approximately four billion US dollars annually (Hoare *et al.* 2008). Given the immense size of these costs, at its thirteenth session the COP encouraged all Parties to "support capacity building, provide technical assistance, facilitate

Table 9.1 Cost required for REDD-plus implementation

	Cost for REDD-plus readiness activities	Cost for REDD-plus project activities	Cost for compensation
Content	• Capacity building for implementing REDD-plus • Establishment cost for forest resources survey such as remote sensing or field survey • Establishment of law and regulation systems such as forest exploitation right • Building of REDD-plus strategies	• Cost for implementing REDD-plus strategies • Implementation of preventive measures for deforestation and forest degradation • Implementation of measures to increase forest carbon stock • Implementation of carbon measurement	• Cost of compensation by implementing preventive measures for deforestation or forest degradation, or measures to promote increase in carbon stock
Cost estimations	• Building monitoring system (0.5–2 million USD for India and Brazil) • Preparation for forest resources survey (50 million USD for 25 countries) • Capacity building (4 billion USD for 40 countries over five years)	• Survey of forest resources (7 million to 17 million USD for 25 countries)	• Opportunity cost to reduce deforestation and forest degradation to half of the current situation (12 billion to 35 billion USD annually) • Opportunity cost for preventing deforestation in eight countries (7 billion USD annually)

Source: Modified Dutschke *et al.* (2008).

Note: USD denotes US dollars.

the transfer of technology . . . and address the institutional needs of development countries" with reference to REDD-plus.

The costs for implementing and evaluating REDD-plus activities depend in part on the type of activity. REDD-plus demonstration activities, or projects, which the thirteenth COP encouraged developing countries to implement, provide us with an indication of what these activities might be. REDD-plus activities common to some of the demonstration activities are patrolling against illegal logging, measures to prevent forest fires, sustainable forest-management activities, and promoting the natural regeneration of forests. Little information on implementation costs is available.

Payment of compensation to those suffering economic losses as a result of the REDD-plus activities (that is, incurring opportunity costs) is also necessary. The opportunity costs may be associated with the opportunities of selling and (or) using timber and non-timber forest products or converting lost forests for agriculture. The opportunity cost to reduce deforestation to half the current level by 2030 has been estimated at between 17 billion and 33 billion US dollars annually (Eliasch 2008).

The relationship between REDD-plus costs and net forest area

The observation that countries seem to experience a common trend in changes in their forests over time led to the formulation of the forest transition theory (Mather 1992) (see Figure 9.4), which has received a lot of attention in relation to REDD-plus. The theory holds that countries with high forest cover will eventually experience high rates of deforestation, while countries with low forest cover will experience low rates of deforestation. The theory also holds that countries that lose much of their forest will eventually begin to restore forest cover as they realize the importance of forest functions.

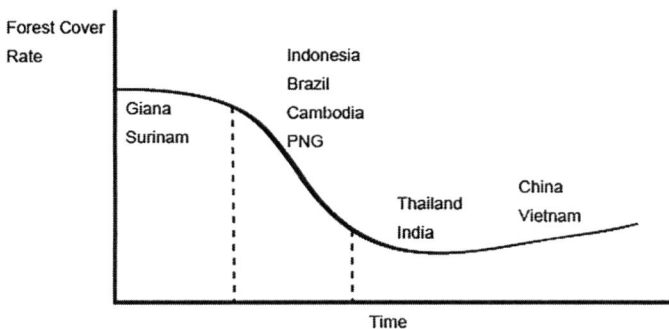

Figure 9.4 Transition of forest-area change in developing countries.

Source: Authors: based on Mather (1992).

Countries can be positioned along the forest transition curve according to the size of their forests (or the percentage of their land that is forested) and the changes taking place to their forest area. For example, there are countries such as Brazil and Indonesia (see Table 9.2) with high forest cover and high rates of forest loss;[1] there are countries such as India, Vietnam, and China with lower percentages of forest cover but increasing forest area (Hyakumura *et al.* 2010; FAO 2010); and there are other countries that are yet to experience rapid deforestation (Chokkalingam *et al.* 2001).

Drivers of deforestation and forest degradation appear to differ according to the size of the national net forest area, and REDD-plus must be tailored accordingly (Wertz-Kanounnikoff and Kongphan-apirak 2009). In countries where deforestation rates are low, market drivers to convert forest lands and to exploit forest resources are not so strong. These tend to be countries where forests are managed according to customary rights and systems, and where legal ownership is not always clear. If market forces strengthen, however, then forest rights are likely to be sought actively and acquired by private investors, resulting in an acceleration of deforestation. In countries with high forest cover and low deforestation rates, REDD-plus financing must thus focus on clarifying forest-use rights and developing sustainable forest-management plans. In countries with high forest cover and high deforestation rates, on the other hand, REDD-plus financing will most likely have to contend with the conversion of large forest tracts for agriculture and timber plantations, and unsustainable logging conducted either legally or illegally. Law enforcement, clarity of user rights, integrated land-use planning, standards to guide and verify sustainable forest management, and transparent governance are the key to REDD-plus success in these countries. Resourcing of forestry departments and law enforcement agencies will also need attention. Even in countries with good laws to support sustainable forest management, forestry departments may be insufficiently resourced to carry out their duties. Finally, in countries where

Table 9.2 Countries with largest annual net loss of forest area (1990–2010)

1990–2000		2000–2010	
Country	*Net area loss (1,000 ha/year)*	*Country*	*Net area loss (1,000 ha/year)*
Brazil	2,890	Brazil	2,642
Indonesia	1,914	Australia	562
Sudan	589	Indonesia	498
Myanmar	435	Nigeria	410
Nigeria	410	Tanzania	403
Tanzania	403	Zimbabwe	327
Mexico	354	Congo	311
Zimbabwe	327	Myanmar	310
Congo	311	Bolivia	290
Argentina	293	Venezuela	288

Source: Modified FAO (2010).

the net forest areas are increasing, REDD-plus can provide additional finance to support existing conservation, land rehabilitation, and plantation policies.

There are two challenges here for REDD-plus financing. First, REDD-plus financing must respond to the underlying and proximate drivers of deforestation in each country, which reflect their net forest area, or where they lie on the forest transition curve. Second, whether equity between countries is important or not will need to be considered. If REDD-plus financing is directed solely at generating and trading emissions offsets, countries with high forest cover and high deforestation rates may act as magnets, whereas countries that are more successful in managing their forests, and countries yet to experience high rates of deforestation, may be less attractive for this finance.

Funding for REDD-plus

The need for a substantial commitment to financing for REDD-plus was highlighted in the Copenhagen Accord which is a non-legally binding document that Parties at the fifteenth COP agreed to "take note of." The Accord (p. 3) reads

> Scaled up, new and additional, predictable and adequate funding as well as improved access shall be provided to developing countries, in accordance with the relevant provisions of the Convention, to enable and support enhanced action on mitigation, including substantial finance to reduce emissions from deforestation and forest degradation.

How the results-based activities are to be financed is yet to be agreed by the COP, and indeed this is one of the most contentious aspects of REDD-plus, with some countries vehemently opposed to carbon markets, and (or) developed countries purchasing carbon credits delivered by REDD-plus to offset their own emissions. Possible sources of REDD-plus financing are public sources, which could be drawn from domestic budgets, new lines of official development assistance (ODA), or redirected existing ODA flows, and private sources, including carbon markets. The options for financing have conventionally been divided into fund-based approaches, market-oriented approaches, and hybrid approaches that combine funds and markets. More recently, the terms "basket" of financing options and "alternative sources" have been introduced.

Under the fund-based approach, international organizations and developed countries would establish and contribute to funds that would compensate REDD-plus activities. Possibilities are funds created by multilateral organizations, regional funds, and bilateral funds under a partnership between one developed and one developing country. One advantage of advance funds is that upfront financing can be provided to meet the costs incurred prior to the actual implementation of the activities. Accordingly, funds are particularly important for the REDD-plus readiness phase, and are particularly relevant to countries that must invest in fundamental forest-governance reforms and institution building. Some point out however, that the amount of funding that is likely be generated through public sources will fall far short of the amounts required for REDD-plus implementation on a large scale (Eliasch 2008).

One initiative that could potentially be tapped for REDD-plus is the Green Climate Fund, which the sixteenth COP agreed to establish as an operating entity of the financial mechanism of the UNFCCC. The Green Climate Fund is intended to support activities in developing countries using thematic windows and, thus, REDD-plus activities could be eligible.

In contrast to the fund-based approach, carbon markets finance REDD-plus by paying for the volume of offsets delivered, measured in tons of CO_2 equivalent, at the market rate. If Parties agree that REDD-plus could be used by developed countries to offset their own CO_2 emissions, markets could provide large sources of financing for both governments and private companies in developed countries. Already, considerable interest in REDD-plus offsets can be detected in the voluntary markets. In their recent global survey of carbon markets, Peters-Stanley *et al.* (2011) found that REDD accounted for the largest share of trade (28 percent) in the over-the-counter trade in 2010.

For markets, sellers must provide assurance that the offsets are real and long-term. Reliable monitoring, reporting, and verification (MRV) systems are necessary. Not only do they provide assurance to markets, but they can also point to ways of enhancing the effectiveness of the REDD-plus activities. The need to develop guidance for MRV systems for all forms of results-based financing, not only markets, is recognized by the COP, which tasked the Subsidiary Body for Scientific and Technological Advice (SBSTA) of the UNFCCC with this responsibility.

There are various risks with market financing for REDD-plus. First, the availability of finance through markets may be highly unpredictable and susceptible to global economic downturns. Second, as carbon is the commodity traded, REDD-plus activities could secure forests for their carbon at the expense of local people's livelihoods, and so on. Third, markets are likely to steer clear of countries with a poor track record in forest governance, and where significant capacity and institution building for REDD-plus is required, resulting in a highly uneven global flow of REDD-plus finances.

The development of voluntary REDD-plus standards for project auditing may suggest a solution to the second risk. The Climate, Community, and Biodiversity Alliance standard, for example, aims to provide assurance that REDD-plus activity will have positive climate, community, and biodiversity outcomes. Standards to deliver REDD-plus offsets to markets do not have to be restricted to the quantification of carbon.

The hybrid approach to financing REDD-plus combines funds with markets. It tailors finance according to needs, whether for REDD-plus readiness activities or to compensate results-based actions. The concept of a basket of financial options has been discussed recently and is similar to the hybrid approach in recognizing both fund and markets, but there is a preference for a mixture of options that developed and developing countries can choose from, according to their interests and needs. This concept is reflected in the Durban Package agreed at the seventeenth COP which states that, "results-based finance provided to developing country Parties that is new, additional and predictable may come from a wide variety of sources, public and private, bilateral and multilateral, including alternative sources" (UNFCCC 2011). The expression "alternative sources" refers to non-market joint mitigation and adaptation approaches that support and strengthen governance, the application of safeguards, and the multiple functions of forests. How these might be developed is yet to be elaborated.

At its sixteenth session in December 2010, the COP reached the Cancun Agreements which include the resolution that REDD-plus should be implemented in phases (UNFCCC 2010). For developing countries to be in a position where they can accurately measure and report their forest sector net emissions reductions, as well as implement activities to achieve these reductions, will take a number of years. A phased approach allows developing countries to focus on preparing themselves for REDD-plus, but it also encourages them to undertake REDD-plus activities, even if they are unable to report the impacts of these on national forest emissions fully; that is, their national MRV systems are not in place. In phase one, developing countries focus on their national strategies or action plans, policies and measures, and capacity building. In phase two, they continue with these activities, but also undertake results-based demonstration activities. In phase three, they implement fully measured, reported, and verified results-based actions. Whether the future negotiations on REDD-plus financing agree on funds, markets, fund-market combinations, baskets, or alternative sources, they need to be informed by the financing requirements of developing countries as they progress through these phases, as well as the likely volume and predictability of finance that each option will deliver.

Existing financing channels

There are now a number of pioneering initiatives providing various forms of public and private finance for REDD-plus (see Table 9.3). The climate negotiations should monitor these closely for possible lessons for a future global funding-framework. Here, we provide a brief description of these financial supports.

Support provided under international initiatives includes the Forest Carbon Partnership Facility (FCPF) established by the World Bank; the United Nations Collaborative Program on Reducing Emissions from Deforestation and Forest Degradation in Developing Countries (UN-REDD), which is a collective endeavor of the Food and Agriculture Organization of the United Nations (FAO), the United Nations Development Program (UNDP), and the United Nations Environment Program (UNEP); and Reducing Deforestation and Forest, and Degradation and Enhancing of. Environmental Services (REDDES) is an initiative of the International Tropical Timber Organization (ITTO). These funds all support REDD-plus readiness activities at national and sub-national levels, as well as sub-national demonstration activities. The FCPF and UN-REDD both support the development

Table 9.3 Existing funding channels for REDD-plus

Type	Title	Administering organization(s)	Details
Funds from international organizations	Forest Carbon Partnership Facility (FCPF)	World Bank	Amount: 170 million USD provided by 11 developed countries. Participation: 37 developing countries. Aim: Supports REDD-plus readiness activities.
	UN-REDD Program	Food and Agriculture Organization of the United Nations (FAO), United Nations Development Program (UNDP), United Nations Environment Program (UNEP)	Amount: 75 million USD provided by developed countries. Participation: 14 developing countries. Aim: Supports REDD-plus readiness activities.
	Reducing Deforestation and Forest Degradation and Enhancing Environmental Services (REDDES)	International Tropical Timber Organization (ITTO)	Amount: 4m. USD (Financial Year 2009) provided by ITTO. Participation: nine developing countries. Aim: Supports MRV, institutional and capacity-building, studies, and demonstration activities.
Bilateral and regional programs	Congo Basin Forest Fund (CBFF)	African Development Bank	Amount: Initial grant of 100 million pounds sterling provided by the United Kingdom and Norway. Participation: Portfolio of 35 projects. Aim: Funds projects to strengthen capacity and reduce poverty and emissions from deforestation.
	Amazon Fund[1]	Brazilian Development Bank	Amount: 116 million USD project approvals. Participation: 20 projects. Aim: Raise donations for investment to preserve the Amazon Biome.

Table 9.3 Continued

Type	Title	Administering organization(s)	Details
	Official Development Assistance	Bilateral aid agencies	Amount: 3,946.6 million USD (indicative interim financing from bilateral sources).[2] Participation: Various developed and developing countries. Aim: Varies between donors but includes readiness, demonstration, and results-based activities.
Private sources Varies on a project-by-project basis	Varies on a project-by-project basis		Amount: 28.4 million tonnes CO2 equivalent REDD offsets traded in the voluntary markets in 2010.[3] Participation: conservation NGOs, private companies, etc. Aim: Conserve forests by generating carbon offsets for voluntary markets or a future compliance market.

Source: Created by the authors.

Note: USD denotes US dollars.
1: http://www.climatefundsupdate.org/listing/amazon-fund
2: Intergovernmental taskforce survey (2010), REDD+ Partnership 2010.
3: Peters-Stanley *et al.* (2011).

of national strategies for REDD-plus readiness, but their approaches are quite different. The FCPF provides countries with funding to develop a basic idea note on how they will prepare for REDD-plus and, if this is accepted, then further funding for development and implementation of a more fully fledged national readiness plan. It will later conduct a trial of performance-based payments for REDD-plus activities in about five countries. The approach of UN-REDD is more direct in that whereas the FCPF provides templates, they provide guidance and an approval process for participating countries to receive funding for the development of their REDD-plus strategies; UN-REDD works directly with governments to draft national joint program documents that identify gaps and actions to fill these. UN-REDD is also active in organizing dialogues and field testing. In contrast, REDDES makes available funding for various types of REDD-plus related activities for which ITTO member countries can apply.

Bilateral and regional funds also take various forms (see Table 9.3). Bilateral funds are mostly directed at readiness activities, but they also include support for demonstration activities. Support for REDD-plus from Norway is unique; it includes funding for results-based policy actions. For example, Norway has concluded a letter of intent with Indonesia under which its support is conditional on Indonesia implementing a two-year suspension of all new concessions for conversion of peat and natural forest. The Amazon Fund and Congo Basin Forest Fund, both regional funds, are somewhat similar to REDDES in that they are financed by developed countries and invite proposals for REDD-plus related activities.

Existing private sources of finances include upfront funding for REDD-plus project development and performance-based finance through the sale of carbon offsets. With the exception of the California Climate Action Registry, the compliance markets do not allow the trading of REDD-plus offsets, whereas the voluntary markets (for example the Chicago Climate Exchange and over-the-counter trade) are more accepting of REDD-plus. Of the various voluntary standards available for the audit of REDD-plus offsets, the Verified Carbon Standard (VCS) is the most popular with project developers (Peters-Stanley *et al.* 2011). The Climate, Community, and Biodiversity (CCB) standards are also popular, but they do not quantify carbon benefits. Several REDD-plus demonstration activities (for example in Ulu Masen, Aceh, Indonesia, and Oddar Meanchey, Cambodia) are thus aiming for dual validation against VCS and the CCB standards (Hyakumura 2009). Some of the initiatives are directed towards future compliance markets, including feasibility studies for REDD-plus projects targeting private firms that are being funded by Japan (Yamagishi 2010; METI 2010, 2011; MOEJ 2011).

Conclusion

We can draw a number of conclusions from the preceding discussion. First, REDD-plus has strong support from developed and developing countries and thus is expected to be included in the future global climate framework. However, while the agreements reached at the sixteenth COP and the seventeenth COP have resolved some of the REDD-plus issues, there are still difficult methodological and political concerns to be addressed. In particular, the issue of how results-based actions will be financed has been a thorn in the side of REDD-plus, but the negotiations have been guided skillfully and they are able to progress by opening up a range of fund, market, and alternative options to be considered further. Fortunately, there is time to find a solution to which Parties can agree. The preferred institutional arrangement for REDD-plus is its inclusion under a new comprehensive climate framework. The framework will not be adopted until 2015 and will come into force on or after 2020. Another option promoted by some Parties is to implement REDD-plus under bilateral cooperation mechanisms. Negotiations on the institutional arrangements for REDD-plus will greatly impact financing for readiness and results-based actions, and thus need to be monitored closely.

Second, REDD-plus could have perverse outcomes, which explains the safeguards, and there is thus a need to ensure that financing for results-based actions is conditional. Well-designed REDD-plus can provide unmatched biodiversity and other co-benefits, which partly explains the strong interest of the voluntary markets in REDD-plus; however, if poorly designed, REDD-plus could result in significant externalities, such as loss of livelihoods and biodiversity. There are various options for ensuring that REDD-plus finance leads to positive climate, community, biodiversity, and other outcomes. A broad global standard could be developed that not only quantifies net carbon emissions, but also evaluates the likely impacts of REDD-plus activities on communities, biodiversity, and so on. Here, much can be learnt from the standards now being used by project developers targeting the voluntary markets. Alternatively, the COP could require Parties to provide credible monitoring and reporting on how they are implementing the safeguards; the SBSTA is now developing guidance for national safeguards information systems.

Third, with respect to the incentive payments, participating governments will have to pay attention to the development of robust institutions to manage the payments and to ensure equitable benefit-sharing. There is a risk that existing institutions will be unable to handle the inflow of REDD-plus finances and that finances will be misdirected or misappropriated. New laws to clarify the ownership of carbon rights may be required. The delivery of benefits to local communities is also important and needs to be considered carefully. Not all communities have well-developed institutions to manage inflows of finance, and investments in health, education, and the creation of sustainable livelihoods may be more appropriate than lump-sum payments to them.

Finally, while deforestation and forest degradation in developing countries require urgent attention, REDD-plus readiness and implementation have to be worked through carefully. Quick-fix solutions will not succeed in countries where deforestation and forest degradation are driven by powerful market forces. High-level political support, multi-sectoral coordination, linkages between the various levels of government, and multi-stakeholder processes are all required to ensure broad commitment to, and ownership of, REDD-plus. REDD-plus results-based activities must be determined on a country-by-country basis and thus respond to the drivers of deforestation, which partly reflect their net forest areas and their positions on the forest transition curve. Financing for readiness activities and results-based actions thus needs to be sufficiently flexible to respond to country needs, while the prospect of payment incentives for results-based activities must be strong enough to encourage governments to commit to REDD-plus.

Note

1 Some parts of some countries may be at different points on the forest transition curve. For example, in Indonesia, Kalimantan appears to be on the first part of the curve in experiencing high rates of forest loss, whereas Java, where forests are better managed, is found further along the curve.

References

Brown D., Seymour, F. and Peskett, L., 2008, "How do we achieve REDD co-benefits and avoid doing harm?" in A. Angelsen (ed.), *Moving Ahead with REDD Issues, Options and Implications*, Bogor, Indonesia, CIFOR: 107–118.

Chokkalingam U., Smith, J. and de Jong, W., 2001, "A conceptual framework for the assessment of tropical secondary forest dynamics and sustainable development potential in Asia," *Journal of Tropical Forest Science*, 13(4): 577–600.

Dutschke M., Wertz-Kanounnikoff, S., Peskett, L., Luttrell, C., Streck, C. and Brown, J., 2008, "How do we match country needs with financing sources?" in A. Angelsen (ed.), *Moving Ahead with REDD Issues, Options and Implications*, Bogor, Indonesia: CIFOR: 41–52.

Eliasch, J., 2008, *Climate Change: Financing Global Forests (the Eliasch Review)*, London: Office of Climate Change, Government of the United Kingdom.

Food and Agriculture Organization (FAO), 2006, *Global Forest Resources Assessment 2005*, FAO.

——, 2010, *Global Forest Resources Assessment 2010*, FAO.

Hoare, A., Legge, T., Nussbaum, R. and Saunders, J., 2008, *Estimating the Cost of Building Capacity in Rainforest Nations to Allow Them to Participate in a Global REDD Mechanism*, UK: Chatham House and ProForest.

Hyakumura, K., 2009, "Biodiversity and global warming: forest certification as a forest conservation measure and REDD," in K. Hayashi (ed.), *Basic Knowledge of Biodiversity and Ecosystem and Economy: Recent Trend in Economy and Business regarding Biodiversity in Simple Terms*, Tokyo: Chuohoki: 245–268 (in Japanese).

Hyakumura, K. and Yokota, Y., 2010, "Institutional arrangement and policies of REDD-plus," *Shinrin kagaku* (Forest Science), 60: 19–23 (in Japanese).

Hyakumura, K., Seki, Y. and Lopez-Casero, F., 2010, "The comparative study on forestation programme in developing countries of Asia: From the view point of securing right of local people," *Ringyo keizai* (Forest Economy), 62(11): 1–20 (in Japanese).

Intergovernmental Panel on Climate Change (IPCC), 2007, *Fourth Assessment Report: Climate Change 2007*, Geneva: IPCC.

Kaufmann, D., Kraay, A. and Mastruzzi, M., 2009, *Governance Matters VIII: Aggregate and Individual Governance Indicators, 1996–2008*, Washington, DC: World Bank Institute.

Khare, A., Scherr, S., Molnar, A. and White, A., 2005, "Forest finance, development cooperation and future options," *Review of European Community and International Environmental Law*, 14(3): 247–254.

Mather, A., 1992, "The forest transition," *Area*, 24(4): 367–379.

Ministry of Economy, Trade and Industry of Japan (METI), 2010, *Result of the second public application for "Promotional Projects for the Prevention of Global Warming by Dissemination of Technology" FY2010* (data on 20 October 2010) http://www.meti.go.jp/information/data/c101020aj.html (Accessed: 7 January 2012) (in Japanese).

——, 2011, *Result of the second public application for "Promotional Projects for the Prevention of Global Warming by Dissemination of Technology" FY2011* (data on 9 August 2011) http://www.meti.go.jp/information/data/c101020aj.html (Accessed: 7 January 2012) (in Japanese).

Ministry of Environment of Japan (MOEJ), 2011, *New Mechanism Express 2011*, August Edition, MOEJ (in Japanese).

Peters-Stanley, M., Hamilton, K., Marcello, T. and Sjardin, M., 2011, *Back to the Future: State of the Voluntary Carbon Markets 2011*, Ecosystem Market Place, Bloomberg New Energy Finance.

Phelps J., Webb, E. and Agrawal, A., 2010, "Does REDD+ threaten to recentralize forest governance?" *Science*, 328(5976): 312–313.

Stern, N., 2006, *The Economics of Climate Change: The Stern Review,* Cambridge: Cambridge University Press.

United Nations Framework Convention on Climate Change (UNFCCC), 2010, *Draft Decision [-/CP.16], Outcome of the Work of the ad hoc Working Group on Long-term Cooperative Action under the Convention*, UNFCCC, 30pp.

United Nations Framework Convention on Climate Change (UNFCCC), 2011, Draft decision [-/CP.17], Outcome of the work of the Ad Hoc Working Group on Long-term Cooperative Action under the Convention, UNFCCC, 55pp.

Wertz-Kanounikoff, S. and Kongphan-apirak, M., 2009, *Emerging REDD+: A preliminary survey of demonstration and readiness activities*, CIFOR Working Paper 46, CIFOR, Bogor, 44pp.

Yamagishi, C., 2010, "Recent tendency of global warming preventive measures: Trend for seeking new reduction mechanism," *Rippo to chosa* (Lawmaking and Research), 308: 67–79 (in Japanese).

10 Evaluation of offset schemes with a laboratory experiment

Keisaku Higashida, Kenta Tanaka, and Shunsuke Managi

Introduction

Among the many types of environmental policies and systems, tradable allowance schemes are considered to be effective; under these systems, the rights to utilize the environment or resources are transacted in the market. One typical scheme is the cap-and-trade scheme which has been considered to reduce the environmental burden efficiently. In fact, many theoretical and empirical studies have proven the effectiveness of tradable allowance schemes (for example see OECD 2000). For example, emission-permits trading schemes have been spreading during the past two decades. The United States first introduced this type of scheme for sulfur dioxide (SO_2) in 1990, and it succeeded in reducing SO_2 emissions by half. After that, other types of emission-permits trading schemes have been introduced in the United States, the European Union, and the United Kingdom. The market scale of the EU-Emissions Trading Scheme (EU-ETS), which began in 2005, is large; the permit price is used as an indicator in other policy measures for climate change.

For fisheries resource management, individual transferable quota (ITQ) schemes with total allowable catch (TAC) have been proven to be better than other kinds of schemes such as input controls, and these schemes have been introduced in several countries such as New Zealand, Australia, and Iceland. Theoretically, a TAC-ITQ scheme may achieve two goals simultaneously: resource management and efficient fisheries. Moreover, ITQ can induce optimal investment behavior by fishers and, accordingly, ITQ may work as a kind of industrial policy (Higashida *et al.* 2011).

For the conservation of biodiversity, a kind of tradable allowance scheme has existed since the 1980s. For example, wetland-mitigation banking schemes (offset schemes) were implemented in the United States: under this scheme, when a developer develops a unit of wetland (a unit of land in which indigenous biodiversity is nurtured), it has to restore and (or) create a unit of wetland that is equivalent to the wetland developed. As a result, the quality of wetlands and indigenous biodiversity are preserved in a certain area.

The whole picture of this scheme is depicted in Figure 10.1. Suppose that there are two developers, Developer A and Developer B, who are developing units of

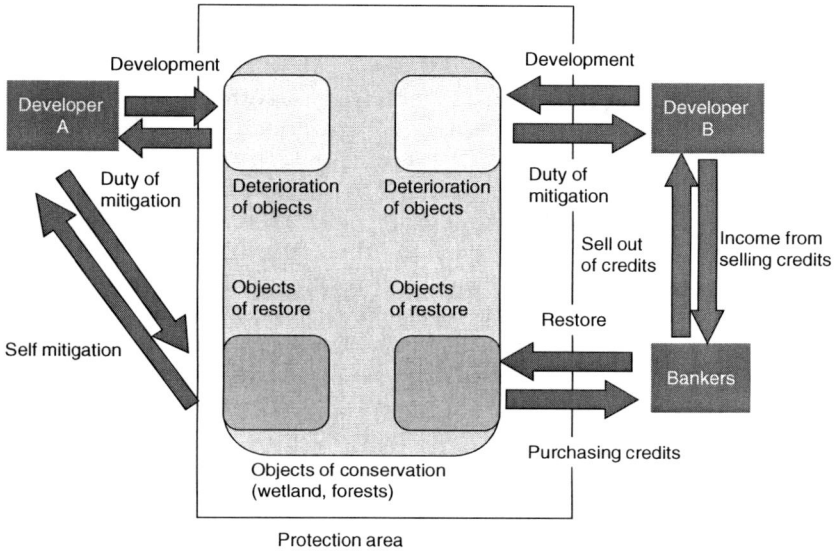

Figure 10.1 Summary of a biodiversity offset scheme.

wetland restricted under an offset scheme. These developers have to restore wet-land in the other points of the area covered by the scheme. The developers have two alternatives. The first approach involves a developer restoring wetland by itself, which is called self-mitigation (see Developer A in Figure 10.1). However, developers do not necessarily have restoration expertise. When this is the case, self-mitigation is very costly, and (or) there is a risk that developers might fail to restore wetlands. The other way is that other players, called bankers, perform the restoration. When a banker restores a unit of wetland it obtains a unit of credit from the authority. Then, if a developer purchases a credit from a banker, the developer is regarded as having restored a unit of wetland. If the cost of restoration by a banker is lower than that by a developer, an offset scheme can achieve the efficient preservation of biodiversity.

Under the United Nations Framework Convention on Climate Change (UNFCCC) an innovative financial mechanism, which includes international transactions of emission permits, has been developed for the prevention of global warming. Global warming gasses targeted by this scheme are international public "bads," which implies that a reduction of a unit of emission in any country has the same effect on global warming.

However, in the case of biodiversity, it is difficult to evaluate two ecosystems in different areas with a certain single measure of evaluation. The reason is as follows. First, indigenous animals and plants vary across areas. Second, even if ecosystems are similar between areas, subjective values in the two areas can differ. For example, the value of ecosystem service of the area in which there are spiritual and (or) cultural symbols can be higher than that of the neighborhood. Thus, it is difficult for authorities to apply tradable allowance schemes directly under the UNFCC for the preservation of biodiversity. However, according to the agreement under the Tenth Conference of The Parties to the Convention on Biological Diversity (COP10), the need for cooperative preservation by many countries has been increasing. In particular, it is believed that an international tradable allowance scheme that covers many areas and (or) countries should be elaborated: a unit of area of ecosystem with high external benefits should be transacted at a high price by establishing an interregional and (or) international financial mechanism, even if the cost of restoration and preservation of such an ecosystem is relatively high.

The purpose of this chapter is to design an offset scheme that can be applied to interregional and (or) international schemes. Then, we evaluate the effectiveness of the scheme with a laboratory experiment.

The remainder of the chapter is organized as follows. The second section describes the outline of our laboratory experiment and the effectiveness of the evaluation of the schemes using laboratory experiments. The third and fourth sections describe the theoretical background and the details of our experiment, respectively. The results of the experiment are examined in the fifth section. The final section provides concluding remarks.

Evaluation of schemes by laboratory experiments

Laboratory experiments are effective for proving theoretical results or for evaluating newly devised schemes. For example, many studies have evaluated emission-permits trading schemes. As noted in the introduction, the Environmental Protection Agency (EPA) of the US government first introduced this type of scheme to reduce the emission of SO_2 and other developed countries followed.[1]

Similar to direct regulations, emission-permits trading schemes are considered a reliable approach for reducing emissions with certainty.[2] Unlike direct regulations, each player observes the permit price and can adjust their own emission flexibly. As a result, the quantities of emissions attributed to two players can be different from each other: the more efficiently a player is able to reduce emissions, the more the player actually reduces emissions. Thus, the total cost of reducing a certain quantity of total emission is minimized.

Although emission-permits trading schemes have been proven to be effective theoretically, it was difficult to evaluate their effectiveness in the real world before and immediately after those schemes were introduced. Thus, a considerable number of studies have been performed on the evaluation of emission-permits trading schemes using laboratory experiments. One of the more controversial points involves how emissions permits are traded. For example, Ledyard and Szakaly-Moore (1994) compared the market efficiency of double auctions and revenue-neutral auctions. Cason and Plott (1996) found the uniform-price auction to be better than the SO_2 emissions allowance trading mechanism of the EPA in the United States. Further, the effects of imperfect competition (Godby 2000; Sturm 2008) and uncertainty (Godby 1997; Cason and Gangadharan 2006) were examined. Additionally, recent studies analyzed more realistic problems specifically. For example, Murphy and Stranlund (2007) tested theoretical predictions concerning compliance behavior in emissions-permits trading schemes. Using a laboratory experiment, Goeree *et al.* (2010) revealed that the grandfathering rule results in dramatic increases in downstream product price.

In the literature, there are two important points in terms of the evaluation of this type of scheme. One is the stability of the permit price and the convergence of prices to the theoretical equilibrium price; the other is the degree of efficiency, that is, whether players trade permits rationally based on their own costs and benefits.

In this chapter, we conduct a laboratory experiment on an offset scheme and evaluate it with the evaluation method of emission-permits trading schemes. Theoretically, when the area for preservation is determined appropriately, offset schemes can achieve the preservation of biodiversity efficiently and, accordingly, maximize social welfare. As far as we know, however, there are few studies that examine offset schemes experimentally. Therefore, it has not yet been verified whether offset schemes, such as wetland-mitigation banking schemes in the United States, work as well as intended.

The basic mechanism of offset schemes is the same as that of emission-permits trading schemes. Therefore, we can apply the results of laboratory experiments on emission-permits trading schemes to the evaluation of offset schemes if the targeted area is small, and if the value of biodiversity can be evaluated by a single measure. However, it is usually difficult to evaluate the values of biodiversity in different areas with a single measure, because the types of biodiversity vary across areas. Therefore, it is important for us to devise an offset scheme that takes into consideration the variety of biodiversity and, accordingly, the difference in external benefits between areas.

In the context of implemented offset schemes, it is easy to evaluate the direct use value in monetary terms using market data. However, it is difficult to measure the indirect use values and non-use values (environmental values). For example, the United Nations Millennium Ecosystem Assessment (2005) classifies ecosystem services into four categories: supporting services, provisioning services, regulating services, and cultural services.[3] Among them, cultural services include the subjective and spiritual benefits that local people enjoy. Therefore, the values of those services cannot be evaluated by a standardized single measure for multiple regions.

It is also difficult to establish a single systematic measure for evaluating regulating services if indigenous plants and animals vary across regions. In some of the existing wetland-mitigation banking schemes, different approaches for setting up relative values for wetlands in different areas have been introduced. For example, in the case of the wetland-mitigation banking scheme in Minnesota, the ratio of the area of developed wetland to the area of restored wetland is determined, which depends on the distance between those two wetlands and the differences in indigenous plants and animals, as well as other conditions. However, no complete standardized method of evaluating ecosystem services in different areas has been established.

In this chapter, we design a scheme that reflects the difference in environmental values in different areas (see Figure 10.2). Assuming that two regions exist and that the environmental value of the unit of wetland in one region is different from that in the other region, we introduce environmental traders into the credit market. Environmental traders buy and sell credits to maximize the sum of the net profits from transactions and the environmental value (net environmental value). We assume that environmental traders are environmental non-government organizations and (or) international bodies. One of the purposes of these organizations and bodies is to preserve and improve the environment, and they know the values of the environment and (or) ecosystem services better than other entrants in the credit market do.[4]

Under our scheme, developers and bankers transact credits within each region. In addition, environmental traders make credit transactions between regions to maximize the net environmental value. Through these transactions, the sum of the consumer and producer surpluses and the environmental value increase.

On the other hand, developers do not take into consideration environmental benefits. They consider only the profits from the development of wetlands and the transactions of credits. Similarly, bankers do not take into consideration environmental benefits. They consider only the cost of restoration and the price of credits. Therefore, when developers and bankers can transact credits across regions, the difference in the environmental values of both regions is not reflected in the credit price. Consequently, the total welfare of both regions is not maximized, and a situation in which wetlands with high environmental values are not sufficiently preserved can occur.[5]

Theoretical background

In this section, we describe the simple theoretical model that corresponds to the offset scheme that we evaluate with a laboratory experiment.

Suppose that there are two regions: regions A and B. An indigenous ecosystem exists in each of the regions. There are also developers and bankers in each region. We assume that: (a) the developers and bankers are price takers in the credit market; (b) the developers in each region are identical with respect to cost and benefit structures; and (c) the bankers in each region are also identical with respect to the cost structure.

In each region, the developers make profits from developing wetlands, although they have to carry out mitigations. We assume that the cost of mitigation (restoration of wetlands) for a developer is much higher than that for a banker and, accordingly, developers choose not to perform self-mitigation but instead carry out mitigation by purchasing credits from bankers. The profit of developer k is given by

$$\pi_{i,k} = R_i x_{i,k} - C_{D,i}(x_{i,k}) - p_i x_{i,k}, \quad C'_{D,i} > 0, \quad C''_{D,i} > 0, \quad i = A, B$$

where R_i, $x_{i,k}$, $C_{D,I}$, and p_i denote the revenue from the development of wetlands in region i, the quantity of wetland developed by developer k in region i, the cost to the development, and the credit price, respectively. The first-order condition (FOC) for the profit maximization is

$$\frac{d\pi_{i,k}}{dx_{i,k}} = R_i - C'_{D,i} - p_i = 0 \tag{10.1}$$

Additionally, the profit of each banker l is given by

$$\Pi_{i,l} = p_i y_{i,l} - C_{S,i}(y_{i,l}), \quad C'_{S,i} > 0, \quad C''_{S,i} > 0, \quad i = A, B$$

where y_i and $C_{S,i}$ denote the quantity of restoration and the restoration cost, respectively. We do not consider carryover and, therefore, bankers sell all of the credits they obtain by restoring wetlands. The FOC is

$$\frac{d\Pi_{i,l}}{dx_{i,k}} = p_i - C'_{S,i}(y_{i,l}) = 0 \tag{10.2}$$

The transaction costs are zero for both intraregional and interregional transactions of credits.[6] Thus, when developers and bankers freely make transactions across regions, the following conditions hold in equilibrium:

$$N_{D,A}x_A + N_{D,B}x_B = N_{S,A}y_A + N_{S,B}y_B \tag{10.3}$$

$$p_A = p_B = p \tag{10.4}$$

where $N_{D,i}$ and $N_{B,i}$ denote the number of developers in region i and the number of bankers in region i, respectively.

Let us now examine the conditions for maximizing world welfare (that is the sum of the profits of developers and bankers and the environmental benefits of both regions given that the total area of wetlands of both regions (that is the sum of the area of wetlands of Region A and that of Region B) is fixed.

We define environmental benefits that are not perceived by bankers and developers as

$$E_i = E_i(N_{S,i}y_i - N_{D,i}x_i + M_{0,i}), \quad E' > 0$$

where $M_{0,i}$ denotes the endowment of wetlands in region i. This is a kind of external benefit and is assumed to be represented in monetary terms. In the case of biodiversity, the external benefits can be classified minutely (see Hassan *et al.* 2005). Because we consider a scenario in which the total area of wetlands of both regions does not vary, the area of wetlands in each region can increase or decrease.

World welfare is defined as

$$GW = N_{D,A}[R_A x_A - C_{D,A}(x_A)] + N_{D,B}[R_B x_B - C_{D,B}(x_B)]$$
$$- N_{S,A} C_{S,A}(y_A) - N_{S,B} C_{S,B}(y_B)$$
$$+ E_A(N_{S,A} y_A - N_{D,A} x_A + M_{0,A})$$
$$+ E_B(N_{S,B} y_B - N_{D,B} x_B + M_{0,B}).$$

Then, the maximization problem can be written as follows:

$$\max_{x,y} \quad GW \quad \text{s.t. } N_{D,A} x_A + N_{D,B} x_B = N_{S,A} y_A + N_{S,B} y_B$$

and the Lagrangian is given by

$$L = N_{D,A}[R_A x_A - C_{D,A}(x_A)] + N_{D,B}[R_B x_B - C_{D,B}(x_B)]$$
$$- N_{S,A} C_{S,A}(y_A) - N_{S,B} C_{S,B}(y_B)$$
$$+ E_A(N_{S,A} y_A - N_{D,A} x_A + M_{0,A}) + E_B(N_{S,B} y_B - N_{D,B} x_B + M_{0,B})$$
$$+ \lambda(N_{D,A} x_A + N_{D,B} x_B = N_{S,A} y_A + N_{S,B} y_B).$$

Thus, the conditions for world welfare maximization are obtained as follows:

$$\frac{dL}{dx_A} = N_{D,A}[(R]_A - C'_{D,A}) - E'_A N_{D,A} - \lambda N_{D,A} = 0 \tag{10.5}$$

$$\frac{dL}{dx_B} = N_{D,B}[(R]_B - C'_{D,B}) - E'_B N_{D,B} - \lambda N_{D,B} = 0 \tag{10.6}$$

$$\frac{dL}{dy_A} = -N_{S,A} C'_{S,A} + E'_A N_{S,A} + \lambda N_{S,A} = 0 \tag{10.7}$$

$$\frac{dL}{dy_B} = -N_{S,B} C'_{S,B} + E'_B N_{S,B} + \lambda N_{S,B} = 0 \tag{10.8}$$

$$\frac{dL}{d\lambda} = 0 \tag{10.9}$$

Can the market equilibrium maximize world welfare? First, suppose that $E''_i = 0$ and $E'_A = E'_B$. In this case, it is clear that the situation in which the conditions from Equation (10.1) through to Equation (10.4) hold is equivalent to the situation in which the conditions from Equation (10.5) through to Equation (10.9) hold. Therefore, if the marginal benefit of a unit of wetland in Region A is the same as that in Region B, and if they are constant, world welfare is maximized when developers and bankers can transact credits across regions.

What if the marginal benefits in both regions are different from each other? For simplicity, we assume that $E''_i = 0$. If $E'_A \neq E'_B$, the situation in which the conditions from Equation (10.1) through to Equation (10.4) hold is not equivalent to the situation in which the conditions from Equation (10.5) through to Equation (10.9) hold. In this case, the credit price for developers and bankers in Region A must be different from that in Region B for world welfare to be maximized. However, if developers and bankers can freely transact credits across regions without any transaction costs, the credit prices in both regions are equal. Therefore, world welfare cannot be maximized.

Then, what kind of scheme should be implemented? We introduce environmental traders into the credit market with the following conditions:

(a) Developers and bankers can transact credits only within each region.
(b) Environmental traders can transact credits both within and across regions. In other words, they buy credits in region i and sell them in region j ($i, j = A, B$, $i \neq j$).
(c) Each environmental trader has to sell all of the credits it bought.
(d) The objective function of an environmental trader is defined as:

$$NE = E_A + E_B - p_A z_A - p_B z_B,$$

where z_i denotes the net quantity of the credit purchase (= the quantity of credits an environmental trader buys in region i minus the quantity of credits it sells in region j).

According to condition (c), $z_A = -z_B$ necessarily holds. In other words, the objective of environmental traders is to maximize the net environmental value that is not captured by developers and bankers.

Suppose that conditions (a) through (d) are satisfied. Then, environmental traders buy credits in Region A and sell them in Region B when $E'_A - p_A - E'_B + p_B > 0$. In contrast, when $E'_A - p_A - E'_B + p_B < 0$, environmental traders buy credits in Region B and sell them in Region A. Because $C''_{D,i} > 0$ and $C''_{S,i} > 0$, the credit price in Region A (Region B) increases (decreases) in the former situation. On the other hand, the credit price in Region B (Region A) increases (decreases) in the latter situation. Consequently,

$$E'_A - p_A - E'_B + p_B = 0$$

holds in equilibrium. It is clear that the situation in which the conditions from Equation (10.1) through to Equation (10.3) and the condition in Equation (10.10) hold is equivalent to the situation in which the conditions from Equation (10.5) through to Equation (10.9) hold. Thus, world welfare maximization can be realized.

Experimental design

The whole picture of the experiment is shown in Figure 10.2. In this experiment, we assumed that there are two regions, Region A and Region B. An offset program

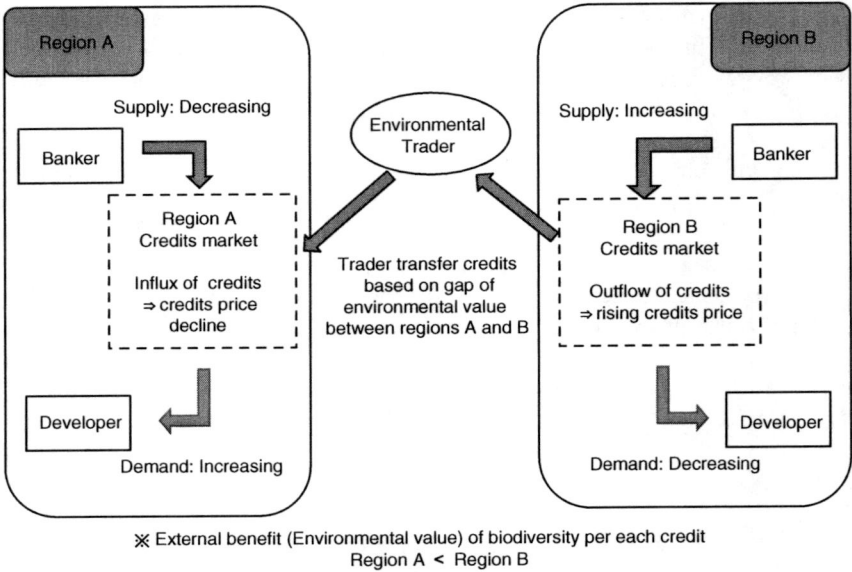

Figure 10.2 Summary of our proposal.

is operated in each of those two regions. There is no difference in environmental values between any pair of two units of wetland within each region, while the environmental value of a unit of wetland in Region A is lower than that in Region B. Except for environmental values, both regions are identical.

As described in the previous section, the quantity of restoration and (or) preservation of wetland in Region B should be greater than that in Region A for the maximization of world welfare. Developers and bankers do not have incentives to take into consideration environmental values when transacting credits and they cannot perform interregional transactions. Instead, environmental traders enter the credit markets and make interregional transactions.

In our experiment, environmental traders were expected to buy credits in the market of Region B and sell them in the market of Region A. In other words, credits move from Region B to Region A such that bankers in Region B have a stronger incentive to restore wetlands and to create credits compared with when no environmental traders enter the credit markets. In contrast, because the inflow of credits decreases the credit price in Region A, bankers in Region A have a weaker incentive to restore wetlands compared with when there are no environmental traders.

We conducted six sessions; 20 subjects participated in each session. Each subject played one of the following three roles: (a) developer; (b) banker; or (c) environmental trader. In each region, four developers and four bankers traded credits only within the market of the region in a computerized double auction.[7] Four environmental traders could make interregional transactions.

The details describing the revenues and costs of developers and bankers are shown in Table 10.1. There are two types of developers in terms of the gains from development in each region. There are also two types of bankers in terms of the cost of restoration in each region.[8] This implies that two subjects are assigned the role of each type of developer in each region, and two subjects are assigned the role of each type of banker in each region. The environmental values in Region A and Region B are four and 16, respectively. Thus, the difference in the environmental values is 12.

If only intra-regional transactions are permitted, the equilibrium price in each region is theoretically 78 through 81. In this case, the volume of credits created is 22. However, when environmental traders enter the credit markets and make interregional transactions to maximize the net environmental value, eight units of credits move from Region B to Region A, and world welfare is maximized. In this case, a difference in the credit prices in both regions is generated: theoretically, the equilibrium price in Region A is 72 through 75, and that in Region B is 84 through 87. If the difference in environmental values is equal to the difference in the credit prices, the environmental values are considered to be internalized into the credit markets.

All subjects were under 30 years old and they were mostly undergraduate and vocational school students. Each subject participated in only one session, and subjects were paid an average of 40 US dollars based on an exchange rate of 77 yen = 1 US dollar. At the beginning of each session, subjects read the instructions for approximately ten minutes. In this experiment, each session included ten periods and the credit markets were open for the entire period, which lasted two minutes. To familiarize subjects with the experiments, we ran two training

Table 10.1 The setting of bankers' costs and developers' revenues

Production · Development unit	Banker (Type 1)	Banker (Type 2)	Developer (Type 1)	Developer (Type 2)
1	60	61	100	99
2	62	63	98	97
3	64	66	95	93
4	68	70	91	89
5	72	75	87	84
6	78	81	81	78
7	84	87	75	72
8	90	94	69	65
9	98	103	61	57
10	108	113	52	47
11	119	125	42	37
12	131	137	31	25

periods before beginning the paying periods. We conducted the experiment using the University of Zurich's Z-tree program (Fischbacher 1999).[9]

The evaluation of the new offset scheme

The basic results

The average variables of all sessions are shown in Figures 10.3 and 10.4.

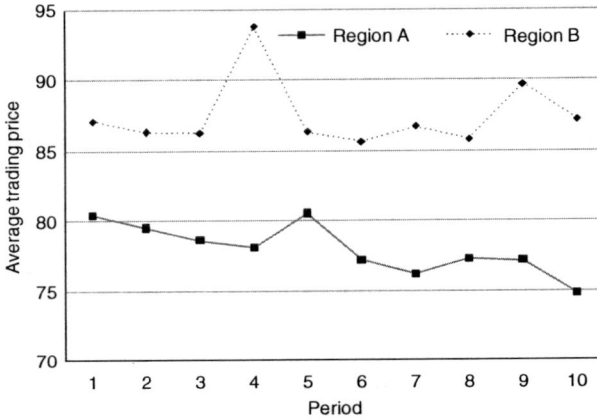

Figure 10.3 The average trading price in each period.

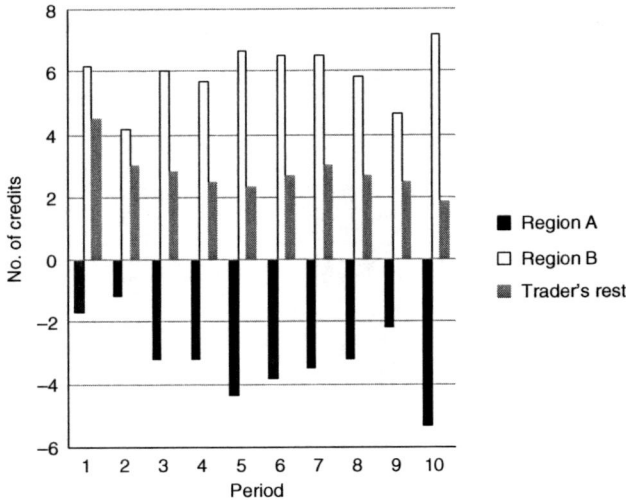

Figure 10.4 The amount of net buying of credits and unsold credits kept in traders' hands (Trader's rest) in each period.

Let us first focus on the credit prices. According to Figure 10.3, the average credit price in Region B was significantly higher than that in Region A. The average price of all periods in Region A was approximately 77, while that in Region B was approximately 87. In general, because the subjects learn the mechanism and their cost and benefit structure in this type of experiment, the average prices tend to converge to theoretical equilibrium prices as they go through the periods. In our experiment, the difference in the credit prices of both regions was continuously approximately ten, and the convergence of the actual transaction prices to the theoretical equilibrium prices was also observed. Although the transaction prices in Region A tended to be higher than the theoretical equilibrium price in the first half, the expected convergence clearly took place. Thus, with regard to the credit prices, the environmental traders functioned as intended, and the optimal credit distribution was achieved.

Let us turn to the behavior of the environmental traders (see Figure 10.4). In all periods, environmental traders bought credits in the market of Region B and sold them in the market of Region A. In the first few periods, traders kept a relatively large quantity of unsold credits in hand. This behavior might have led to the high credit prices in Region A in the first half. However, the quantities of unsold credits tended to decrease as time passed.[10]

Efficiency

Next, we examine whether credits were created efficiently. Efficiency means that credits are created so that world welfare is maximized. Specifically, we use the method of efficiency evaluation developed by Ledyard and Szakaly-Moore (1994). The index is given by

$$Efficiency = \frac{Experimental\ Payoff - Status\ Quo\ Value}{Maximum\ (Efficient)\ Payoff - Status\ Quo\ Value}$$

where the status quo values are equal to zero in this case, because no credits exist and, therefore, no development takes place. Experimental payoff is the sum of the payoffs of all subjects realized in each period, and maximum payoff is the sum of the payoffs of all subjects in each period when the optimal creation and transactions of credits were realized. In this experiment, to include the environmental values of preserved biodiversity, we use the following definition for experimental payoff:

$$Experimental\ Payoff = \sum \pi_{trader} + \sum \pi_{developer,A} + \sum \pi_{developer,B}$$
$$+ \sum \pi_{banker,A} + \sum \pi_{banker,B},$$

where π denotes payoffs. Note that $\sum \pi_{trader}$ includes not only the profits or loss from credit transactions, but also the environmental values. These values are denoted by $E'_A \times S_A + E'_B \times S_B$, where E'_i and S_i denote the marginal environmental value of wetland in region i and the net quantity of created wetlands

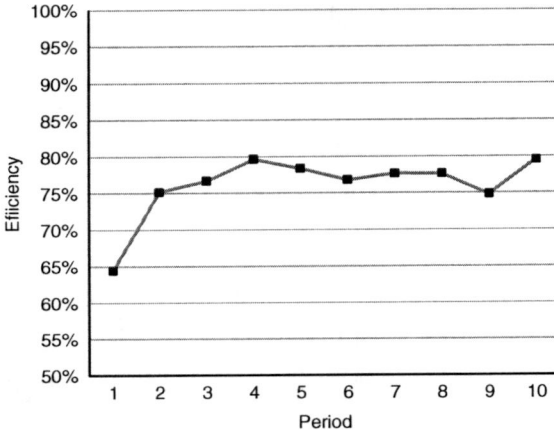

Figure 10.5 The average efficiency in each period.

(= the units of created wetland minus the units of developed wetland) in region i, respectively. In the present setting of the experiment, $E'_A = 4$ and $E'_B = 16$.

The results of the efficiency analysis are shown in Figure 10.5. The numbers indicate the average efficiency of all sessions in each period. The efficiency is low in the first period, while it approaches 80 percent from the second through the last periods. This result indicates that subjects learned their objectives and that their behavior came close to the optimal behavior as they went through periods. The efficiency was stable after it approached the 80 percent point. Although efficiency generally approaches 90 percent in the case of a simple double-auction mechanism, our experiment is more complicated than other simple experiments are. Thus, the scheme with environmental traders functioned well.

The effect of entry of environmental traders

In the previous section, we showed that the scheme with environmental traders may function well. However, it is not enough to clarify whether efficiency under the scheme with environmental traders is significantly greater than that without environmental traders. Thus, in this subsection, we examine the effect of the entry of the environmental traders using Ordinary Least Squares (OLS) estimation. We estimate the following equation for each region:

$$abs(TCA - CA_{m,r}) = c + \beta_1 TTA_{m,r} + \beta_2 TTA^2_{m,r} + \beta_3 Rest_{m,r}$$
$$+ \beta_4 abs(PG_{m,r}) + \beta_5 VTP_{m,r} + \varepsilon,$$

where

m: the index of sessions
r: the index of periods
abs: absolute value
TCA: theoretically optimal quantity of credits
CA: the quantity of credits created in the experiment
TTA: the quantity of transactions made by environmental traders
Rest: the quantity of credits that are unsold and kept in traders' hands
PG: the difference between the theoretical equilibrium price and the actual
   transaction price in the experiment
VTP: the variance of transaction prices.

We also estimate the following equation, which includes the variables of both regions:

$$abs(TCA - CA_{m,r}) = c + \beta_6 TTA_{m,r} + \beta_7 TTA_{m,r}^2 + \beta_8 Rest_{m,r}$$
$$+ \beta_9 abs(PG_{m,r}^A) + \beta_{10} abs(PG_{m,r}^B)$$
$$+ \beta_{11} VTP_{m,r}^A + \beta_{12} VTP_{m,r}^B + \varepsilon$$

The dependent variables in both estimating equations are the absolute value of the difference between the theoretical optimal quantity of credits and the quantity of credits created in the experiment, which we refer to as an inefficient creation of credits.

We adopt five independent variables. When environmental traders behave rationally, their transactions achieve the optimal creation of credits in both regions. Therefore, β_1 and β_6 are expected to be negative. However, environmental traders may behave speculatively. Thus, we also include TTA^2 as an independent variable. β_2 and β_7 are expected to be positive, because we extract the effect of excess transactions caused by speculation.

In the present setting of the environmental values, when environmental traders keep credits in their hands they lose. Moreover, in such a case, the optimal creation of credits cannot be achieved. Thus, β_3 and β_8 are expected to be positive. It is likely that the greater the difference between the theoretical equilibrium price and the actual trading price, the greater the difference between the theoretically optimal quantity of credits and the quantity of credits created in the experiment will be. Therefore, β_4, β_9, and β_{10} are expected to be positive. The price difference is either positive or negative. Therefore, if the theoretical equilibrium price minus the actual trading price was taken as a sample, it would be impossible to estimate the effect of the price difference precisely. Thus, we adopt absolute values. Finally, the variance of credit prices may influence the decision-making of the bankers and developers in each region. In general, when the variance is larger, the behavior of bankers and developers is not rational.

The results of the estimation are shown in Table 10.2. The quantity of transactions made by environmental traders (*TTA*) negatively affected the inefficient creation of credits in all estimations, while TTA^2 positively affected the inefficient creation of credits. In our experiment, the average quantity of transactions

Table 10.2 Factor analysis of excess development and underdevelopment

All		Region A		Region B	
TTA	−1.201***	TTAa	−0.581*	TTAb	−1.654***
TTA^2	0.018***	TTAa2	0.025**	TTAb2	0.042***
Rest	−0.610**	Rest	−0.575**	Rest	−0.27
PGa	0.336***	PGa	−0.303***	PGb	0.311*
PGb	0.608***	VTPa	0.001**	VTPb	−0.001*
VTPa	−0.001*	C	5.203*	C	19.86***
VTPb	−0.001***				
C	26.072***				
R^2	0.382		0.414		0.1

Note: *Significant at 10% level, **Significant at 5%, ***Significant at 1% level.

in a period is 15.03 in Region A and 18.13 in Region B. Taking into consideration those quantities of transactions, the effect of *TTA* dominates that of TTA^2 and, accordingly, the transactions by environmental traders are considered to reduce the inefficiency on the creation of credits. In other words, as expected in terms of theory, the entry of environmental traders improves the efficiency of the preservation of wetlands in both regions.

PG^A and PG^B positively influenced the inefficient creation of credits. This implies that the greater the gap is between the theoretical optimal price and the actual price, the greater the inefficiency of preservation of wetlands is. This is intuitive, because the subjects can behave rationally as price takers. It is observed that PG^A in the estimation of Region A does not have any significant influence on the inefficient creation of credits in Region A. One possible reason is that it takes some time for environmental traders to buy credits in Region B and sell them in Region A. Therefore, it is possible that a certain quantity of credits is created before the inflow of credits takes place. In our experiment, this trend can be observed in particular in the first few periods, which may influence the insignificant result or the price gap.

In the case of this type of experiment, it is often observed that when the variance of the price is greater, the preservation of environmental values is less efficient. In contrast, according to our estimation, the coefficient of *VTP* is significantly negative, which implies that the variance reduces the inefficiency. However, the value of the coefficient of *VTP* is extremely small compared with other coefficients. Therefore, the effect of the variance seems to be trivial. A possible reason is that the credit transactions across regions can cause large changes in credit prices, which implies that a greater variance in the credit price does not necessarily lead to greater inefficiency in credit creation.

Conclusion

In this chapter, we designed an offset scheme – which can be applied to interregional and (or) international schemes – for internalizing the value of biodiversity into the market. The specific feature is the entry of environmental traders who can evaluate the difference in the environmental values of different regions. Then, we evaluated the effectiveness of the scheme with a laboratory experiment.

In the laboratory experiment, we assumed the existence of two regions. The environmental traders could perform credit transactions across regions, while the developers and bankers could only perform transactions within a region. Then, we examined whether the traders could improve the efficient creation of credits in both regions. According to the results, the traders could not make a sufficient quantity of transactions to achieve this efficiency in the first few periods. However, as they went through periods, their behavior became rational because of the learning-by-doing effect; therefore, the efficiency of credit creation was improved. The estimation results also revealed that environmental traders took into consideration the difference in the environmental values of both regions when they made transactions. We should note a caveat. In our laboratory experiment, we excluded important factors that could affect the function of credit markets, for example asymmetric information on credit transactions and the uncertainty and (or) expectation of changes in the rules of offset schemes and (or) demand and supply conditions. However, other types of tradable allowance schemes, such as tradable emission permits and ITQ also have these problems. In those schemes, mechanisms for tackling these problems were elaborated and have already been introduced. For example, banking systems[11] make it possible for traders to reduce the risk caused by unexpected events of holding insufficient permits, which leads to stability in the permit price.

Recently, methods for evaluating specific environmental values have been developed by economists and ecologists, and their accuracy has been rapidly improving (for example, Kassar and Lasserre 2004). Thus, it has become realistic to introduce an offset scheme with environmental traders. Moreover, the evaluation of ecosystem services and the preservation of biodiversity are going to proceed at full swing. When taking into account that other kinds of tradable allowance schemes have been introduced already and are functioning well, it becomes feasible that offset schemes will be able to contribute to the preservation of biodiversity. However, the value of biodiversity varies across regions due to factors such as the variety of indigenous animals and plants across regions. Even if those animals and plants are the same, environmental values may be different among regions because external benefits can differ. For example, a unit of wetland not only nurtures indigenous plants but it also influences the environment of adjacent areas.

Thus, mechanisms that are specific to the offset schemes of ecosystem services, such as the entry of environmental traders, should be elaborated. Through such mechanisms, the total welfare of more than two regions can be maximized by taking into consideration the differences in environmental values. When considering the long-term preservation of biodiversity, the introduction of a financial mechanism that covers more than one region is indispensable. In this respect, the evaluation of offset schemes and the elaboration of complementary mechanisms will become important.

Notes

1 The Japanese government has been operating a type of emission-permit trading scheme on a trial basis since 2005. The Tokyo prefectural government also began a similar scheme in 2010.
2 We assume that complementary schemes of monitoring and punishment work well.
3 See Hassan *et al.* (2005) for the details.
4 In terms of the reflection of cultural services on the market price of credits, a citizens' group can be an environmental trader when it is well-versed in the local culture and the traditional value of ecosystems for local people.
5 When focusing on one region, the sum of consumer and producer surpluses and the environmental value can be maximized only by transactions between developers and bankers, because the total quantity of wetlands is fixed and because economic benefits are generated.
6 In reality, the transaction costs are likely to be close to zero, because, unlike goods, developers and bankers do not need to transport credits.
7 The double-auction mechanism has been proven to be better than other types of transaction mechanisms in terms of efficiency and the stability of credit prices when the sellers and buyers are clearly differentiated. See for example Smith (1962).
8 We introduce two types of bankers and developers to increase the stability of credit prices.
9 These sessions were conducted at the Tohoku University in September and October 2010.
10 In this experiment, we exclude the banking and borrowing of credits across periods – which implies that traders cannot keep these unsold credits in the next period.
11 Banking systems have the authority to allow credits that were not used in the current season to be used in the following season. Credit banking can create a cushion that will prevent price spikes and can hedge uncertainty in the credit price.

References

Cason, T. N. and Gangadharan, L., 2006, "Emissions variability in tradable permit markets with imperfect enforcement and banking," *Journal of Economic Behavior and Organization*, 61(2): 199–216.
Cason, T. N. and Plott, C. R., 1996, "EPA's new emission trading mechanism: A laboratory evaluation," *Journal of Environmental Economics and Management*, 30(2): 133–160.
Fishbacher, U., 2007, "z-Tree–Zurich toolbox for readymade economic experiments," *Experimental Economics*, 10(2): 171–178.
Godby, R. W., 1997, "Emissions trading with shares and coupons when control over discharges is uncertain," *Journal of Environmental Economics and Management*, 32(3): 359–381.
——, 2000, "Market power and emissions trading: Theory and laboratory results," *Pacific Economic Review*, 5(3): 349–363.
Goeree, J., Palmer, K., Holt, C., Shobe, W. and Burtraw, D., 2010, "An experimental study of auctions versus grandfathering to assign pollution permits," *Journal of the European Economic Association*, 8(2–3): 514–525.
Hassan, R., Scholes, R., and Ash, N., 2005, *Ecosystems and Human Well-being: Current State and Trends (Millennium Ecosystem Assessment)*, Washington, DC: Island Press.
Higashida, K. and Tanaka, 2011, *Evolutional Mechanism of Biodiversity Offset: Proposition Based on Evaluation by Experiment Approach*, Discussion paper.

Kassar, I. and Lasserre, P., 2004, "Species preservation and biodiversity value: A real options approach," *Journal of Environmental Economics and Management,* 48(2): 857–879.

Ledyard, J. O. and Szakaly-Moore, K., 1994, "Designing organizations for trading pollution rights," *Journal of Economic Behavior and Organization*, 25(2): 167–196.

Millennium Ecosystem Assessment, 2005, *Ecosystems and Human Well-being: Current State and Trends*, Washington, DC: Island Press.

Murphy, J. and Stranlund, J., 2007, "A laboratory investigation of compliance behavior under tradable emissions rights: Implication for targeted enforcement," *Journal of Environmental Economics and Management*, 53(2): 196–212.

OECD, 2000, *Implementing Domestic Tradable Permits for Environmental Protection*, OECD Publishing, 10.1787/9789264181182-en.

Smith, V. L., 1962, "An experimental study of competitive market behavior," *Journal of Political Economy*, 70(2): 111–137.

Sturm, B., 2008, "Market power in emissions trading ruled by a multiple unit double auction: Further experimental evidence," *Environmental and Resource Economics*, 40(4): 467–487.

TEEB, 2010, *The Economics of Ecosystems and Biodiversity–Ecological and Economic Foundations,* P. Kumar (ed.), London: Earthscan.

Wråke, M., Myers, E., Burtraw, D., Mandell, S. and Holt, C., 2010, "Opportunity cost for free allocations of emissions permits: An experimental analysis," *Environmental and Resource Economics*, 46(3): 331–336.

Part IV

Economic analysis on ecosystem restoration and resource management

11 Project portfolio analysis of global ecosystem restoration

Kei Kabaya and Shunsuke Managi

Introduction

We humans substantially benefit from goods and services provided by ecosystems (MA 2005), and we cannot survive without these ecosystem services. Nevertheless, those have been perceived as externalities; their genuine values have not been evaluated in the market and in policies, rather, ecosystems are recognized as valueless. As a result, development and over-extraction providing easily visible economic profits have been prioritized thus far, and so ecosystems have been continuously degraded at the global level.

In order to break through this situation, attempts have been made recently to measure the economic values of ecosystem services. For instance, Costanza *et al.* (1997) estimated that the total global economic value of such services would reach 33 trillion US dollars per annum (as of 1994), although the methodology has been evaluated as problematic. Also, The Economics of Ecosystem and Biodiversity (hereafter TEEB) initiated by the United Nations Environmental Program (UNEP) and the European Union has analyzed the cost of policy inaction, estimating that ecosystem services equivalent to 2.0 to 4.5 trillion US dollars would be lost over a 50-year period (TEEB 2008). From these studies, the enormous economic value of ecosystem services has been gradually recognized.

If ecosystems have economic value, then their restoration is environmentally and also socio-economically important. In fact, cost–benefit analysis on the restoration of nine ecosystems demonstrated that benefits could exceed costs in every case (TEEB 2009); this implies that ecosystem restoration is economically efficient when their economic value is taken into account.

Meanwhile, considering that financial constraints regarding ecosystem restoration do exist even these days (James *et al.* 2001), it will be difficult to allocate sufficient financial resources to restore all ecosystems. In such a situation, we need to investigate which ecosystems we should primarily target as investment destinations. In this respect, the above-mentioned study did not clearly mention what was a priority, although it demonstrated that grassland ecosystems recorded the highest cost–benefit ratio, which may imply that single-pole investment in grassland would be the most economically efficient. However, uncertainty about ecosystem restoration projects will involve risks.

Assuming that international public agencies or environmental bodies would decide investment allocation to ecosystem restoration upon recognition of any socio-economic benefits, as well as economic efficiency, what kind of analytical framework would be appropriate? Considering the above risks of single-pole investments, the viewpoint of a portfolio needs to be applied for analysis. Simultaneously, the unique definitions of return and risk that are different from modern portfolio theory will be required, taking into account the uniqueness of ecosystem restoration.

This chapter will propose one analytical framework for the purpose of contributing to this decision-making. Here, project portfolio[1] which is usually applied for decision-making regarding resource extraction and product invention will be the basis for this analysis. We will define return as the cost–benefit ratio of the ecosystem restoration, and risk as the possibility of over-budget cost.[2] Furthermore, one hypothetical ecosystem restoration model will be constructed and the relevant data will be obtained, thereby enabling analysis of investment efficiency at a global level.

Few studies based on the above analytical framework have been seen; however, a thought experiment in this analysis is expected to contribute to further discussion on investment into ecosystem restoration. Also, socio-economic perspectives applied herein may have a unique value, because the traditional approach for resource allocation to ecosystems has mainly focused on the judgmental standard based just upon ecological importance (for example Murdoch *et al.* 2007; Underwood *et al.* 2008).

Ecosystem restoration project portfolio model

As ecosystem restoration requires a certain timeframe, a hypothetical dynamic model regarding ecosystem restoration will be constructed herein. Although this model applies a number of unique assumptions due to the complexity of ecosystem restoration and the existence of few mathematical models thereof, reflection of ecological theories regarding restoration paths as well as effects of stable states and disturbance will be a basis for adding further sophistication to the model.

First of all, an ecosystem is assumed to restore in logistic fashion, applying the existing theoretical studies on ecosystem restoration (for example Suding *et al.* 2004). Additionally, the following points will be taken into consideration:

- Ecosystem states at the beginning of restoration will be different between projects
- The edge effect regarding habitat areas will affect the restoration process
- Characteristics of restoring works will have impacts on restoration speed.

Based on these assumptions, the following model will be framed:

$$s_t = \{1 + b \times \exp(-a \times c \times t)\}^{-1} \tag{11.1}$$

$$(0 \leq s_t \leq 1, 1 \leq t \leq T, b > 0, 0 < a \leq 1, 0 < c \leq 1),$$

where s is an ecosystem state,[3] t denotes year, b stands for a coefficient affecting an original state, a and c are coefficients indicating the effects of restoration areas and restoration costs, respectively, and T denotes total project duration; each variable should fulfill each condition within the bracket. This equation implies that a better original state as well as bigger restoration areas and higher costs will accelerate ecosystem restoration.

As a next step, disturbance and stable states under ecosystem restoration projects are modeled. Restored ecosystems will be affected externally, with a possibility of degradation, while an ecosystem has several certain stable states into which it will settle when disturbed (Beisner *et al.* 2003; Scheffer *et al.* 2001). As several stable states exist, it may be possible to restore the ecosystem from a lower stable state to a higher one (Suding *et al.* 2004). On the contrary, a fall from a higher stable state to a lower one can be assumed to occur in the case of a strong disturbance, and a further complete collapse of an ecosystem can also be expected. Here, two states of e_n are presumed (e_1: lower stable state; e_2: higher stable state), and s^* (ecosystem state at the time of disturbance) and s' (ecosystem state restoration after disturbance) are modeled as follows, considering disturbance magnitude d_m (d_1: weak disturbance; d_2: middle disturbance; d_3: strong disturbance), which could occur at the end of the year:

$$s_{t^*}^* = e_1 \ (0 < e_1 \leq 1) \tag{11.2}$$

s.t. $s_{t^*} > e_2$ and $d_m (m = 3)$, or $e_1 < s_{t^*} < e_2$ and $d_m (m = 2)$

$$s_{t^*}^* = e_2 \ (0 < e_2 \leq 1)$$

s.t. $s_t{}^* > e_2$ and $d_m (m = 2)$

$$s_{t^*}^* = 0$$

s.t. $s_{t^*} < e_1$ and $d_m (m = 1, 2, 3)$, or $e_1 < s_{t^*} < e_2$ and $d_m (m = 3)$

$$s'_t = \left\{ 1 + b \times \exp\left[a \times c \times \left(-t + t^* + \frac{\ln\left(\frac{e_n - 1}{b \times e_n}\right)}{a \times c} \right) \right] \right\}^{-1} \ (t_1^* < t \leq T)$$

s.t. $s_{t^*}^* \neq 0,$ $\tag{11.3}$

where s_{t^*} and $s_{t^*}^*$ denote an ecosystem state just before a disturbance and that at the time of a disturbance, respectively; and t^* and t_1^* are the year of the latest disturbance and that of first disturbance, respectively. From the above, a restored ecosystem would retreat to each stable state depending on the state at the moment and the disturbance magnitude when a disturbance occurs, and it then would proceed once more on a similar pathway to the one before the disturbance, as long as it does not fully collapse. Meanwhile, a weak disturbance d_1 is assumed to have no effects on an ecosystem state, except the case of one below a lower stable state e_1. Some sample paths of restoration based on Equation (11.1) through Equation (11.3) are shown in Figure 11.1.

On the basis of this ecosystem restoration model, the benefit B_t at time t as well as the restoration cost C_t are set as follows, assuming that benefits are obtained

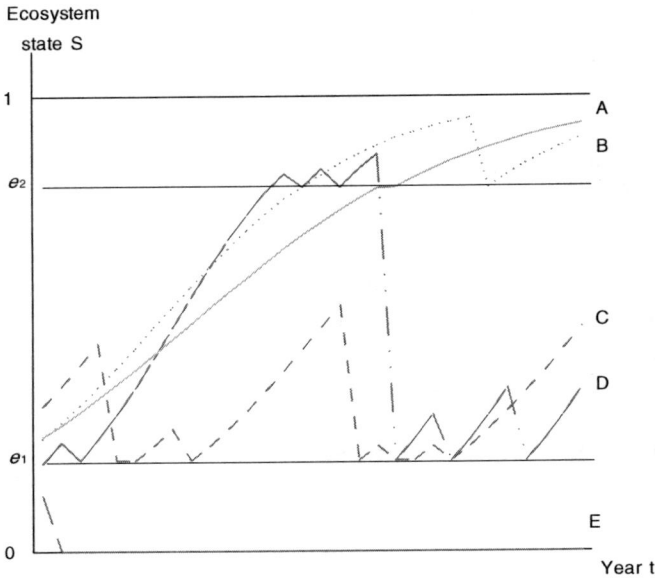

Figure 11.1 Examples of ecosystem restoration path.

Note: A is the example of restoration with almost no effects of disturbance, and B is the example of one time disturbance after passing e_2. C implies drop from e_2 to e_1 due to strong disturbance, and D indicates unprogressed restoration owing to frequent middle disturbance. E is the extreme case of ecosystem collapse because of disturbance at the beginning.

at the end of the year, costs are earmarked at the beginning of the year, and the discount rate r is constant over the period of T:

$$B_t = B_o \times A \times S_t \times (1 + r)^{-t} \tag{11.4}$$

$$C_t = C_o \times A \times (1 - S_t) \times (1 + r)^{-(t-1)} \quad (0 \le S_t < e_2)$$

$$C_t = 0 \quad (e_2 \le S_t \le 1),$$

where S indicates an ecosystem state which adopts s, s^*, and s', depending on the above conditions. From the above, benefits and costs are dependent on the original benefit B_o, the original cost C_o, the restoration area A, and the ecosystem state S; benefits are increasing and costs are decreasing as an ecosystem state is improving. Here, no restoration activities are assumed to be required in the case of an ecosystem state beyond a higher stable state e_2.

The return (cost–benefit ratio R) and the risk (over-budget possibility V) of ecosystem restoration projects with those benefits and costs can be expressed as follows:

$$R = \sum_{t=1}^{T} B_t / \sum_{t=1}^{T} C_t$$

$$V = P \left[\sum_{t=1}^{T} (C_t^* - C_t) < 0 \right]$$

s.t. $C_t^* = \alpha \times C_o \times A \times (1 - s_t) \times (1 + r)^{-(t-1)} (\alpha \geq 1, 0 \leq s_t < e_2)$

$C_t^* = 0 \quad (e_2 \leq s_t \leq 1),$

where $P(\cdot)$ is probability, C^* denotes restoration budgets, and α indicates a budgetary coefficient. As a restoration budget estimates the costs without considering the disturbance, but with an expectation of a higher than required amount, over-budget or under-budget may occur. In reality, it will be difficult to predict disturbance intensity at the time of calculating a restoration budget, and it may be reasonable to estimate higher budgets just in case. As such, the return R_p and the risk V_p of an ecosystem restoration project portfolio are framed as follows, similar to Equations (11.6) and (11.7):

$$R_p = \sum_{i,j} \sum_{t=1}^{T} (B_{i,j,t} w_{i,j}) / \sum_{i,j} \sum_{t=1}^{T} (C_{i,j,t} w_{i,j}) \quad (w_{i,j} = 0, 1)$$

$$V_p = P \left[\sum_{i,j} \sum_{t=1}^{T} \{((C_{i,j,t}^* - C_{i,j,t}) w_{i,j})\} < 0 \right] \quad (w_{i,j} = 0, 1)$$

where subscripts i and j denote ecosystem and region, respectively, and w indicates project implementation; 0 indicates no implementation, while 1 indicates the opposite. Note that Equation (11.1) through Equation (11.7) will take different figures depending on ecosystem i and region j.

Data and simulation setting

The database which systematically organizes economic benefits from ecosystems and restoration costs is extremely limited; specifically, it is impossible at this stage to obtain the one which indicates ecosystem restoration costs as well as the economic benefits derived from that restoration, simultaneously. Hence, one database each on economic benefits from ecosystems and on restoration costs is referred to herein, acknowledging that there is no direct relationship between them.

As for the former, the necessary data for simulation are obtained from the Cost of Policy Inaction (COPI) (Ten Brink *et al.* 2009). Here, 528 pieces of data on GDP-adjusted annual economic benefit per unit hectare in Euro in 2007 are converted into US dollars in 2007, and categorized into the ecosystems and regions indicated below. Subsequently, their average value and standard deviation are calculated.[4] As for the latter, the data indicating the investment amount and project areas are extracted from the Global Restoration Network database; 69 pieces of data are obtained. Annual restoration costs in US dollars in 2007 are estimated based on each project's duration first, and then annual restoration costs per unit hectare are calculated by dividing that value by project areas. Similar to economic benefits, the average value and the standard deviation of restoration costs as well as project areas and duration of each category are estimated.

In consideration of the amount of data and the similarity of habitats, six ecosystems and six regions are nominated as categories. Among those, combinations which hold more than three pieces of data on all of the economic benefits, restoration costs, project areas, and project duration are nine in total (see Table 11.1).[5, 6] Basic statistics based on the above classifications are shown in Table 11.2.

In the simulation, random digits regarding economic benefits, restoration costs, project areas and project duration will be generated based on each average and standard deviation on the assumption of a log-normal distribution. Among those random digits, the former three are directly substituted with B_o, C_o, and A, respectively, while the latter three are applied for c, a, and b, respectively with some manipulations.[7]

Among five drivers of change listed in MA (2005: 16), climate change, invasive alien species, and pollution are selected as disturbance factors for restored ecosystems.[8] As driver impact on biodiversity over the last century as well as its current trend are indicated qualitatively here, they will be quantified as follows: the former will be evaluated with an ordinal scale from 1 (small) to 4 (large), and the latter will be expressed as a change rate from −50 percent to 100 percent.[9]

Table 11.1 Combination between ecosystems and regions

	Ecosystem	*Habitats included*	*Region*	*Abbreviation*
1	Coastal areas	Seagrass beds, estuaries, and tidal wetlands	North America	Coast-NA
2	Mangrove*	Mangrove	Asia	Mangr-AS
3	Freshwater	Rivers, lakes, and wetlands	North America	Fresh-NA
4	Freshwater	Rivers, lakes, and wetlands	Asia	Fresh-AS
5	Freshwater	Rivers, lakes, and wetlands	Europe	Fresh-EU
6	Temperate and boreal forests	Temperate forests, Mediterranean forests, and boreal forests	North America	Tempf-NA
7	Temperate and boreal forests	Temperate and boreal forests	Europe	Tempf-EU
8	Tropical forests	Tropical forests	Latin America	Tropf-LA
9	Grasslands and drylands	Shrublands, savanna, grasslands, and desert	Africa	Grass-AF

Note: As sufficient data can be collected, mangrove is analyzed independently of coastal areas, while coral reefs are excluded from this analysis due to data deficiency.

Table 11.2 Basis statistics

	Item	Coast-NA	Mangr-AS	Fresh-NA	Fresh-AS	Fresh-EU	Tempf-NA	Tempf-EU	Tropf-LA	Grass-AF
Annual economic benefit ($/ha)	No. of data	3	21	9	34	11	19	24	36	14
	Min	117	36	641	0	140	0	0	0	0
	Max	1,012	629,993	53,502	41,387	36,963	5,857	8,032	23,638	481
	Average	498	93,842	16,474	6,155	5,692	383	1,960	2,181	73
	SD	462	198,389	17,796	9,889	11,165	1,335	2,311	4,573	156
Annual restoration cost ($/ha)	No. of data	17	4	7*	3*	4	5	3	3	3
	Min	2	204	725	9	166	1	9	236	1
	Max	180,079	8,225	21,114	1,696	20,586	3,825	852	2,217	1
	Average	34,288	4,225	7,523	810	13,278	1,076	490	988	1
	SD	46,626	4,213	7,455	847	9,071	1,586	401	1,037	0
Restoration area (ha)	No. of data	17	4	7*	3*	4	5	3	3	3
	Min	1	5	2	800	2	4	365	16	160,000
	Max	896,100	20,000	10,400	465,114	4,000	607,050	33,000	630	3,112,366
	Average	93,385	5,151	1,530	157,193	1,553	122,381	12,022	249	1,153,701
	SD	263,271	9,902	3,912	266,678	1,931	270,944	18,205	333	1,696,314
Restoration duration (Year)	No. of data	17	4	7*	3*	4	5	3	3	3
	Min	0.5	3.42	1.50	3.00	2.17	4.00	4.00	3.00	3.00
	Max	35.25	9.92	19.00	12.08	15.00	22.00	15.08	8.00	15.58
	Average	9.23	6.96	9.08	7.56	6.54	10.82	8.08	4.92	8.00
	SD	10.29	2.68	6.80	4.54	5.76	6.75	6.09	2.70	6.68

Note: *One data item is excluded for each due to its relatively gigantic annual restoration cost.

Furthermore, the impacts will be divided by 10, which would be the basis of disturbance possibility, and such possibility is assumed to change at the rate of current trends linearly during the period T, while the change rates may be affected by a due to the edge effects of ecosystem areas. Based on the above assumptions, disturbance possibility $P(D_{f,i,j,t})$ will be modeled as follows:

$$P(D_t) = \frac{P(D_o) \times \left(1 + \frac{u \times t}{T}\right)}{1 + a}, \tag{11.5}$$

where $P(D_o)$ denotes an original disturbance possibility, and u implies change rates during the period T. In the simulation, these disturbances are assumed to occur when $P(D_t)$ exceeds random digits x. When three disturbance factors make their effects simultaneously, it can be defined as a strong disturbance d_3, when two middle disturbanced_2, and when one weak disturbance d_1. Each factor in Equation (11.5) will take different figures depending on ecosystems, regions, and drivers of change. However, taking into account each characteristics of disturbance, all ecosystems and regions are assumed to share the same random digits for climate change, those for invasive alien species will be shared among the same ecosystem, and the same region will share the same random digits for pollution.[10]

As for other relevant values in the simulation, $T = 30$ and $r = 4.45$ percent will be applied in the base case.[11] An ecosystem stable state e_n is completely unknown, thus 0.2 and 0.8 will be arbitrarily substituted for e_1 and e_2, respectively. Last, 20 percent higher costs will be estimated for the restoration budget, that is $\alpha = 1.2$.

Simulation results

First, 10,000 iteration results based on the Monte Carlo method are indicated individually in Table 11.3. By and large, higher cost–benefit ratios were marked in the developing region, while the developed regions, inter alia North America, recorded lower cost–benefit ratios. As over-budget possibilities were higher in the regions of Asia and North America, investment in Asia can be said to be high risk–high return, and in North America high risk–low return. Terrestrial ecosystems experienced less frequent disturbances; on the contrary, coastal areas and freshwater ecosystems were vulnerable and recorded higher project failure probabilities. All in all, Grass-AF was the most economically efficient investment destination from the perspective of a cost–benefit ratio, while Tropf-LA was the safest investment destination in terms of an over-budget possibility. The result of a high cost–benefit ratio of grasslands is compatible with that of TEEB (2009), as mentioned above. However, the values themselves are greatly different due to the use of different data and models. Therefore, the result could not be concluded as being highly reliable, but this would be highly meaningful to indicate a tendency of a cost–benefit ratio of ecosystem restoration projects.

Based on each cost–benefit ratio and over-budget possibility, these ecosystem restoration projects are classified as in Figure 11.2. As their classification is based on relative comparisons, the boundaries thereof are set as the medians

Table 11.3 The simulation results in the base case

Item	Coast-NA	Mangr-AS	Fresh-NA	Fresh-AS	Fresh-EU	Tempf-NA	Tempf-EU	Tropf-LA	Grass-AF
Economic benefits (1,000$)	291,314	3,825,482	173,673	6,129,072	56,467	451,426	207,047	4,921	683,192
Restoration costs (1,000$)	18,068,804	114,271	67,312	753,923	129,719	561,750	28,326	1,076	4,487
Restoration budget (1,000$)	13,688,903	105,157	53,620	613,432	113,903	575,833	30,273	1,207	4,872
Disturbance frequency	6.277	3.306	5.560	5.671	5.756	1.474	1.496	0.967	2.684
Project failure possibility (%)	62.02	43.34	52.89	53.65	53.10	25.24	25.87	22.91	35.28
Cost–benefit ratio	0.013	25.923	1.685	6.556	0.281	0.894	5.246	4.362	99.418
Over-budget possibility (%)	60.01	56.38	60.61	59.65	58.08	46.46	43.41	32.03	53.81

Note: the italic font denotes the median value of each item.

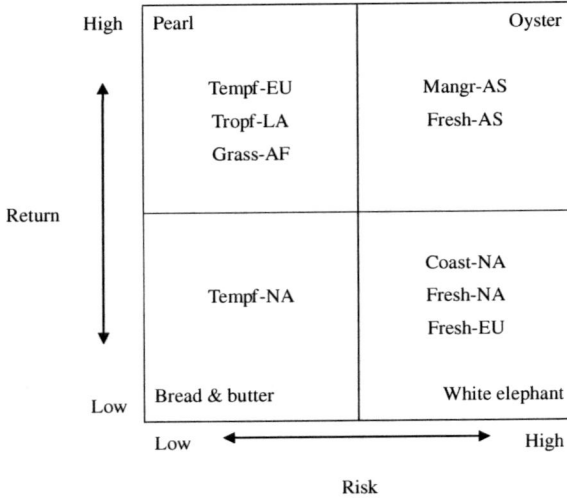

Figure 11.2 Project portfolio matrix.

Note: the boundary standard of return is cost–benefit ratio = 4.362, and that of risk is over-budget ratio = 56.38%. The figures which fall on the boundary will be incorporated into the upper or righter matrix.

of nine ecosystem restoration projects. The characteristics of each box are as follows: "Pearl": low risk–high return, "Oyster": high risk–high return, "Bread & butter": low risk–low return, "White elephant": high risk–low return (Matheson and Matheson 1998). According to this project-portfolio management, the Pearl should be highly prioritized, the White elephant should be excluded or reconsidered, and the rest should be allocated to the Oyster and the Bread & butter. Thus, Tempf-EU, Tropf-LA, and Grass-AF are the prime candidates for an investment destination.

To verify the stability of this classification in the base case, similar iterations under the following three different scenarios were performed:

- Scenario 1: shortening period T from 30 years to 10, followed by decreasing the discount rate to 4.04 percent;[12]
- Scenario 2: transforming e_1 and e_2 to 0.1 and 0.9, respectively;
- Scenario 3: changing the budgetary coefficient from 1.2 to the ones that reduce the gaps, with the actual costs based on the iteration results in the base case.

As a result of each of the iterations, both the cost–benefit ratio and the over-budget possibility in each scenario are differentiated from the base case; Scenario 1 recorded smaller values for both, while Scenario 2 increased only the over-budget possibilities; likewise, larger figures of the latter were shown in Scenario 3.

Table 11.4 The estimation results of ecosystem restoration project portfolio

Portfolio			Cost–benefit ratio		Over-budget possibility (%)
			95% confidence interval		
Tempf-EU	Tropf-LA		5.567	(5.384–5.749)	42.96
Tempf-EU	Grass-AF		34.536	(32.177–36.895)	44.24
Tropf-LA	Grass-AF		78.324	(73.137–83.511)	49.38
Tempf-EU	Tropf-LA	Grass-AF	33.758	(30.336–34.714)	43.57

Although the median values as boundary standards were different in each scenario, the same classifications as in the base case were obtained in every scenario. In brief, the results of this classification are concluded not to be affected by the project period, discount rate, stable state, and budgetary coefficient.

As certain stability for the classification of ecosystem restoration projects could be affirmed as above, the project portfolio will be analyzed with this base case as a next step. First, four portfolios consisting of some of the three ecosystem restoration projects classified as Pearl were investigated by means of 10,000 iterations, resulting in the highest return for the one consisting of Tropf-LA and Grass-AF, and the lowest risk for the one consisting of Tempf-EU and Tropf-LA (see Table 11.4). As for the former portfolio, the risk was improved from a single investment on Grass-AF, although the return was decreased; similarly, the latter portfolio demonstrated the highly improved return from a single investment on Tropf-LA, although the risk worsened. Then, the ecosystem restoration projects classified as Oyster and Bread & butter were added to the above two project portfolios, and a similar iteration was performed. As a result, no improvement either in return or risk was marked.

Aspiration level of return and acceptance level of risk depend on individual value judgments, but if maximization of a cost–benefit ratio under the constraint of an over-budget ratio set lower than 50 percent is pursued, the project portfolio which consisted of Tropf-LA and Grass-AF can be concluded as being the most preferable investment destination here. In brief, these simulation results demonstrated that the application of project-portfolio viewpoints would improve investment efficiency on ecosystem restoration in some aspects, when defining return as a cost–benefit ratio and risk as an over-budget possibility and using the above model as well as data.

Discussion on the other investment allocation criteria

Traditionally, ecological importance has tended to be underlined as an investment standard in ecosystem restoration, but the question is how efficient will such an investment standard be from the socio-economic viewpoints applied in this chapter. In addition, it must be asked how effective will the investment standard of a cost–benefit ratio be, as well as looking at the over-budget possibility, in an ecological sense.

Using the above ecosystem restoration model framework, the other investment-allocation criteria, namely the designated protected area ratio and the endangered species ratio, will be analyzed herein. As for the former, the regional ecosystems which mark a relatively lower designated protected area ratio, thereby being preferentially invested are first specified based on the database provided by Chape *et al.* (2008: 36–75) (see Table 11.5). Then, such ecosystem restoration projects will be incorporated into the project portfolio in the order from that with the smallest protected area ratio up to that with its median value. As a result of iteration on four portfolios using the base case ecosystem restoration model, an extremely low cost–benefit ratio and a high over-budget possibility were illustrated (see Table 11.6).

Next, an endangered species ratio is calculated based on the Red List provided by the International Union fpr Conservation of Nature (IUCN) so as to specify the regional ecosystems which hold a relatively higher endangered species ratio (see Table 11.5). Similar to the above, such ecosystem restoration projects will be incorporated into the project portfolio in descending order from those with the highest endangered species ratio down to those with a median value. As a result of iterations on five portfolios, the cost–benefit ratio was higher than that of the case of a protected area ratio on average, but an over-budget possibility remained higher (see Table 11.6). In sum, investment allocation based on a protected area ratio as well as an endangered species ratio will lose the opportunities to gain a higher socio-economic cost–benefit ratio and thus reduce the over-budget possibility.

From a different perspective, the contribution to the protected area ratio as well as the endangered species ratio of the most preferable project portfolio prioritized by the above simulation results, namely the one consisting of Tropf-LA and Grass-AF, will be investigated. Here, an improvement in the protected area ratio, as well as in the endangered species ratio per unit of investment of 1,000 USD, are applied as standard measures for effectiveness. As for the former, all restored areas are assumed to be protected hereafter, and the total percentage increase in the protected area ratio is perceived as an improvement; as for the latter, the number of endangered species per hectare multiplied by the project areas is used as a contribution to the endangered species.

As a result of the above evaluations, Table 11.7 was formulated. Simultaneously, assessment was made of contributions from the other project portfolios consisting of two regional ecosystems, which are prioritized by each investment standard, that is, the protected area ratio and the endangered species ratio. In conclusion, the project portfolio based on the investment standard of a cost–benefit ratio as well as an over-budget possibility as analyzed in this chapter recorded the best scores (see Table 11.7). From these analyses, investment-allocation criteria based on a cost–benefit ratio and an over-budget possibility can be said to be superior not only from socio-economic aspects, but also on an ecological cost-effectiveness basis.

Table 11.5 Designated protected area ratio and endangered species ratio

Item	Coast-NA	Mangr-AS	Fresh-NA	Fresh-AS	Fresh-EU	Tempf-NA	Tempf-EU	Tropf-LA	Grass-AF
Total area (1,000 ha)	1,774,000	5,833	290,000	210,000	20,000	181,400	812,600	822,900	2,581,100
Protecte area (1,000 ha)	21,213	–	92,800	63,000	5,600	21,200	147,800	209,170	279,329
Protected area ratio (%)	1.20	–	32.00	30.00	28.00	11.69	18.19	25.42	10.82
Red listed species	985	132	1,484	5,966	1,654	915	1,027	9,737	3,756
Endangered speices*	91	18	*264*	1,189	577	58	167	2,944	482
Endangered species ratio (%)	9.24	13.64	17.79	19.93	34.89	6.34	*16.26*	30.24	12.83
Endangered species ratio per unit	0.00005	0.00309	*0.00091*	0.00566	0.02885	0.00032	0.00021	0.00358	0.00019

Note: the italic text denotes the median value of each item. In terms of protected area and its ratio, as the data on protected area of mangrove is unobtainable, no italic font is present herein: the median value of the protected area ratio will be 21.80%.

Table 11.6 Cost–benefit ratio and over-budget possibility of the project portfolios based on each investment standard

Investment standard	Portfolio				Cost–benefit ratio		Over-budget possibility (%)
						95% confidence interval	
Protected areas ratio	Coast-NA				0.014	(0.014–0.015)	60.58
	Coast-NA	Grass-AF			1.055	(0.932–1.179)	60.56
	Coast-NA	Grass-AF	Tempf-NA		0.649	(0.582–0.716)	59.66
	Coast-NA	Grass-AF	Tempf-NA	Tempf-EU	0.750	(0.688–0.811)	59.57
Endangered species ratio	Fresh-EU				0.267	(0.251–0.282)	57.75
	Fresh-EU	Tropf-LA			0.493	(0.459–0.528)	57.54
	Fresh-EU	Tropf-LA	Fresh-AS		4.832	(4.553–5.111)	58.28
	Fresh-EU	Tropf-LA	Fresh-AS	Fresh-NA	4.724	(4.456–4.991)	59.52
	Fresh-EU	Tropf-LA	Fresh-AS	Fresh-NA	Tempf-EU	5.030	(4.769–5.290)	59.23

Table 11.7 Contribution to protected area ratio and endangered species ratio of the project portfolios prioritized in each investment standard

Investment standard	Cost–benefit ratio and over-budget possibility		Protected area ratio		Endangered species ratio	
Project portfolio	Tropf-LA	Grass-AF	Coast-NA	Grass-AF	Fresh-EU	Tropf-LA
Restoration costs (1,000 USD)	5,562		18,073,291		130,795	
Improvement in protected area ratio per 1,000 USD (%)	8.931×10^{-7}		0.003×10^{-7}		0.054×10^{-7}	
Contribution to the number of endangered species per 1,000 USD	387.699×10^{-7}		0.121×10^{-7}		3.498×10^{-7}	

Note: USD denotes US dollars.

Conclusion

This analysis, which defined return as a cost–benefit ratio and risk as an over-budget possibility, and which applied the above hypothetical model and limited data, demonstrated that international public agencies or environmental bodies could make better decisions on investment allocation to ecosystem restoration by applying project-portfolio viewpoints. The traditional approach for investment allocation regarding ecosystems has certainly emphasized ecological importance, but it also demonstrated that investment judgment from socio-economic aspects is superior to others in terms of not only socio-economic but also ecological cost effectiveness. In brief, considering socio-economic benefits and applying project portfolios are crucial for investment judgment on ecosystem restoration.

However, the values of these results significantly depend on the model and data, as stated above, which may imply that these values are not necessarily correct all the time. Hence, sophistication of the model and improvement of the data will be required for a further study. Moreover, the data should fulfill the condition of simultaneous indication of restoration costs and economic benefits derived from that restoration for a more accurate simulation. Last, although no budget constraints were set up herein, portfolio analysis should be performed with restoration costs and the project areas based on them when applying this in reality.

This chapter demonstrated the importance of socio-economic viewpoints in investment allocation to ecosystem restoration and revealed the necessity for sophistication of the model as well as enlargement of the data, which will contribute further to discussions relevant to investment in ecosystem restoration.

Notes

1 It has similar objectives to modern portfolio theory in the field of finance, namely return maximization as well as risk avoidance, however, several differences can be observed as follows:

- As historical data on the variance of the rate of return rarely exist, it is not necessarily used as a risk (alternatively, the probability of project failure can be used, for example);
- Likewise, correlation of it can be replaced by regional similarity or future operational costs (Moriarty 2001);
- Whether investing in a project or not it is on an all-or-nothing basis.

2 It will be a substantial issue in project management that costs exceed budgets, which can also be applied in ecosystem restoration projects. Such a risk is studied in similar studies (for example Yoe 2001).

3 A pristine state is recognized as the most desirable state herein, whereas the state not providing any conventional functions is perceived as the worst state. Based on this understanding, the former is assumed to generate the maximum economic benefits, while the latter is presumed not to provide any of them.

4 As the data on total economic value of ecosystem services are extremely limited, classification of ecosystem services for evaluation will not be considered herein.

5 Three pieces of data are not necessarily sufficient, but further stringent constraints will drastically reduce the combinations fulfilling the conditions. Thus, these criteria are set herein.

6 No sufficient dataset existed in the region of Oceania.

7 As for c and a, each random digit is transformed into a probability value in each normal distribution so as to fulfill the conditions of $0 < c \leq 1$ and $0 < a \leq 1$, respectively. Likewise, b is transformed into a probability value in a normal distribution and further multiplied by 10 for the purpose of increasing the probability of a higher effect of b of more than 1, thereby avoiding overestimation of original states of ecosystems. The reason why project duration can be interpreted as the coefficient affecting original states is that longer project durations can imply more degraded ecosystem states.

8 In general, restored ecosystems are conserved thereafter, so that habitat change and over exploitation will be excluded.

9 As for the ecosystems differently classified between this analysis and MA (2005), average values of those of relevant ecosystems will be calculated. That is, the figures for mangroves will be calculated as the average values of tropical forests and coastal areas; the average values of subcategories of drylands will be utilized for the figures of grasslands and drylands; and a similar interpretation will be applied for temperate and boreal forests.

10 This presumption would generate correlation between ecosystem restoration projects in terms of economic benefits and restoration costs.

11 This assumes the interest rate of the Treasury Bond in the United States on 31 December 2007.

12 This assumes the interest rate of the Treasury Note for 10 years in the United States on 31 December 2007.

References

Beisner, B. E., Haydon, D. T. and Cuddington, K., 2003, "Alternative stable states in ecology", *Frontiers in Ecology and the Environment*, 1(7): 376–382.

Chan-Lau, J. A., 2010, "Fat tails and their (un)happy endings: Correlation bias and its implications for systemic risk and prudential regulation", *IMF Working Paper*, WP/11/82, IMF.

Chape, S., Spalding, M. and Jenkins, M. D., 2008, *The World's Protected Areas*, prepared by the UNEP World Conservation Monitoring Centre, University of California Press, Berkeley, USA.

Costanza, R., d'Arge, R., de Groot, R., Farber, S., Grasso, M., Hannon, B., Limburg, K., Naeem, S., O'Neill, R. V., Paruelo, J., Raskin, R. G., Sutton, P. and van den Belt, M., 1997, "The value of the world's ecosystem services and natural capital", *Nature*, 387: 253–260.

Global Restoration Network: http://www.globalrestorationnetwork.org/database/ (Accessed: 6 February 2012).

James, A., Gaston, K. J. and Balmford, A., 2001, "Can we afford to conserve biodiversity?", *BioScience*, 51(1): 43–52.

Matheson, D. and Matheson, J., 1998, *The Smart Organization: Creating Value through Strategic R&D*, Boston, MA: Harvard Business School Press.

Millennium Ecosystem Assessment (MA), 2005, *Ecosystems and Human Well-being: Synthesis*, Washington DC: Island Press.

Moriarty, N., 2001, "Portfolio risk reduction: optimizing selection of resource projects by application of financial industry techniques", *Exploration Geophysics*, 32: 352–356.

Murdoch, W. W., Polasky, S., Wilson, K. A., Possingham, H. P., Kareiva, P. and Shaw, R., 2007, "Maximizing return on investment in conservation", *Biological Conservation*, 139(3–4): 375–388.

Scheffer M., Carpenter, S., Foley, J. A., Folke, C. and Walker, B., 2001, "Catastrophic shifts in ecosystems", *Nature*, 413: 591–596.

Suding, K. N., Gross, K. L. and Houseman, G. R., 2004, "Alternative states and positive feedbacks in restoration ecology", *Trends in Ecology and Evolution*, 19(1): 46–53.

Ten Brink, P., Bassi, S., Gantioler, S., Kettunen, M., Rayment, M., Foo, V., Bräuer, I., Gerdes, H., Stupak, N., Braat, L., Markandya, A., Chiabai, A., Nunes, P., ten Brink, B. and van Oorschot, M., 2009, *Further Developing Assumptions on Monetary Valuation of Biodiversity Cost Of Policy Inaction (COPI)*, Institute for European Environmental Policy (IEEP).

The Economics of Ecosystems and Biodiversity (TEEB), 2008, *Interim Report*, TEEB.

——, 2009, *TEEB Climate Issues Update*, TEEB.

The IUCN Red List of Threatened Species, http://www.iucnredlist.org/ (Accessed: 6 February 2012).

Underwood, E. C., Shaw, M. R., Wilson, K. A., Kareiva, P., Klausmeyer, K. R., McBride, M. F., Bode, M., Morrison, S. A., Hoekstra, J. M. and Possingham, H. P., 2008, "Protecting biodiversity when money matters: maximizing return on investment", *PLOS ONE*, 3(1): e1515.

Yoe, C., in association with Planning and Management Consultants, Ltd., 2001, *Ecosystem Restoration Cost Risk Assessment*, IWR Report 02-R-1.

12 Diversity on fisheries

Price volatilities in the Japanese market

Kentaka Aruga and Shunsuke Managi

Introduction

Japan has the world's sixth largest Exclusive Economic Zone (EEZ). The level of biodiversity and ecosystem service in this zone is very large and is a habitat for around 14 percent of all the world's marine organisms (MAFF 2010a).

The FAO (2010) states that stocks of the top ten highly captured marine species from the fisheries industry are fully exploited or are overexploited, and it is believed that some of the fish species in the Japanese EEZ are already overexploited. It is becoming more important for the Japanese government to implement sustainable fisheries management to help protect the stocks in marine resource areas, and to secure the level of biodiversity for its EEZ. If the Japanese fishing industry continues to exploit the fish resources from its EEZ, the fish stocks in the area could be depleted.

The decline in fish stocks in the EEZ area will raise fish prices because more effort will be needed to catch the fish and this will increase the fishing cost. Due to this increased cost of fishing and the difficulty of supplying adequate fish to the market, there may be excess demand and this in turn increases the market price. On the other hand, when fish prices are too low, the cost of fishing could become higher than the expected profit from selling the fish. If this is the case, it may be better for the producers to reduce their catch-effort or stop fishing altogether. This is because fish producers often face a high fixed cost, which includes the cost of maintaining boats and fishing gear and so on. Thus, the market price has a large impact on fish production, affecting the catch-effort of the producers. If prices of certain fish have particularly high volatilities, it would be difficult for producers to set the optimal level of effort to maximize their profits. This implies that producers facing markets with high price volatilities will be involved in high price risks. If the producers want to stabilize their profits it might be preferable for them to change their target from fish with high price volatilities to lower ones. However, little is known about which types of fish have high price volatilities among the various types of fish species consumed in Japan. There are several studies comparing price volatilities for different commodities in the literature of energy economics (for example Plourde and Watkins 1998; Regnier 2007), but there has not been much analysis using prices for fish species.

   This study intends to bridge this gap, by comparing and identifying the price volatilities among various fish species consumed in Japan to provide a better understanding of the risk in fish prices. Japan has been one of the world's largest fish-importing countries and hence its impact on the world seafood market is significant. Understanding the Japanese fish market will be helpful, not only for Japanese consumers to satisfy their current consumption level continuously, but it will also be valuable for fish-exporting countries. One huge difference in the Japanese fish market compared to other countries is that in Japan there are more varieties of fish traded in the market because the Japanese have historically consumed various types of fish (Wessells and Wilen 1994). In particular, we compare the price volatilities among the following 12 major fish species consumed in Japan: *maguro* (tuna), *aji* (horse mackerel), *iwashi* (sardines), *karei* (flounder), *saba* (mackerel),*sanma* (saury), *buri* (yellowtail), *ika* (cuttlefish/squid), *tako* (octopus), *ebi* (prawns), *asari* (short-necked clams) and *hotate* (scallops). The retail price for the following highly populated cities is used for the comparison: Tokyo (special wards, 8.95), Yokohama (3.69), Osaka (2.67), Nagoya (2.26), Sapporo (1.91), Kobe (1.54), Kyoto (1.47), Fukuoka (1.46), Hiroshima (1.17), Sendai (1.05) (parentheses show the population in millions) (Statistics Bureau 2011).
   In the following section, we discuss the data for fish prices used in the study and details of the characteristics of the fish consumption in the ten cities listed above. The third section explains the methods used for comparing the price volatilities. Finally, in the fourth section, we reveal the results and conclusions of the study.

The fish prices used in the study

The monthly price is used for every fish price and all prices are obtained from the retail price survey conducted by the Japanese Statistics Bureau. The period covered in this study is from January 1996 to December 2009. The price for *maguro* is either the price of 100 grams of yellowfin or bigeyed tuna, sliced for sashimi use. For *aji*, *iwashi*, *karei*, *saba*, and *sanma* the prices are for whole fish measuring more than 15 cm, more than 12 cm, between 25 and 35 cm, and more than 25 cm in length.
   The price of *buri* is for 100 grams of sliced *buri*. The price of *ika* is for 100 grams of Japanese Flying Squid. The *tako* price is for 100 grams of boiled octopus. The *ebi* price represents the price for 100 grams of imported frozen, packaged, and headless shrimp that are 8–10 cm long. The price of *asari* is for 100 grams of short-necked clams with shell. Finally, the *hotate* price is for 100 grams of boiled and shelled scallops.
   The fish price obtained from the Japanese retail price survey covered about 70 cities whose population was more than 150,000, or the capitals of the 47 subnational jurisdictions. In this study we used fish prices of the cities with the top ten highest populations: Tokyo, Yokohama, Osaka, Nagoya, Sapporo, Kobe, Kyoto,

Fukuoka, Hiroshima, and Sendai. These cities are often segregated into east and west regions by their geographic locations, and the east and west regions often have different cultures and food preferences.[1] Among the cities used in the study Tokyo, Yokohama, Sapporo and Sendai belong to the east region; Osaka, Kobe, Kyoto, Fukuoka and Hiroshima are located in the west region; and Nagoya falls in middle of these regions.

Table 12.1 shows the average annual fish consumption per household for the cities used in the study. From the average quantity of fish consumed for different cities, the table illustrates that *maguro* and *hotate* are consumed more in cities located in the east region, such as Tokyo and Yokohama. On the other hand, *saba*, *tako*, and *ebi* are consumed more in the cities in the west region, such as Osaka, Kyoto, and Kobe. Consumption for the other fish seems to be also affected by the geographic location of the city. For instance, *aji* and *iwashi* are consumed more in Fukuoka, which is located on the southwest edge of Japan. *Karei*, *sanma*, and *ika* are consumed the most in Sapporo, a city on the northeast edge of Japan. Finally, *buri* and *asari* do not have much of a location difference in the average quantity consumed among the cities.

Table 12.1 also illustrates the differences in the average quantity consumed per household among different types of fish. Comparing the national average annual fish consumption per household for expense and quantity, the table suggests that for an average household the expense and quantity consumed for *maguro*, *ika*, *buri*, and *ebi* are larger than those for *iwashi*, *tako*, *hotate*, and *asari*. Thus, there seem to be differences in the consumption among different types of fish in Japan.

It is likely that this consumption behavior is reflected in the fish price as well. Figure 12.1 presents the retail fish prices of Tokyo, Osaka, and Nagoya for the 12 different fish included in this study.[2] The figure suggests that changes in prices within the year seem to be severe for the *aji*, *iwashi*, *karei*, *saba*, *sanma*, *buri*, and *ika*, which is related to the seasonality in these fisheries. Changes in other fish prices appear to be smaller within the year, and these prices are affected more by differences in demand and supply of a certain year, rather than by the seasonality factor within the year. Overall, the figure shows that none of the fish have the same patterns of price movement and this indicates that every fish has different demand and supply factors, thus creating different market fundamentals. As shown in Table 12.1, it is arguable that differences in the way the Japanese households in different regions consume these fish are also a factor causing these fish prices to move in different patterns. The figure also reflects the different price movements among different cities; this is likely to be related to the geographic location of the cities. In the figure, Tokyo and Osaka represent the market prices for the east and west regions, with Nagoya also included in the figure because this market often shares the consumption behavior of both of the regions. It seems that differences in the price movements among the three cities are not as severe as are the differences among different fish species and, to some extent, these three cities share their price information.

Table 12.1 Average annual fish consumption per household in Japan (2008–10)

	Maguro		Aji		Iwashi		Karei		Saba		Samma	
	Expense	Quantity	Expense	Quantity	Expense	Quantity	Expense	Quantity	Expense	Quantity	Expense	Quantity
National	5,765	2,471	1,500	1,558	613	764	1,540	1,284	1,141	1,328	1,372	2,171
Tokyo	8,724	3,176	1,495	1,386	514	572	1,085	738	864	896	1,310	1,753
Yokohama	8,758	3,165	1,670	1,697	583	832	1,117	757	966	1,117	1,401	1,992
Osaka	4,270	1,896	1,169	1,127	607	668	1,868	1,415	1,277	1,402	1,085	1,495
Nagoya	8,363	3,193	1,121	892	610	694	1,078	754	1,094	1,125	1,253	1,736
Sapporo	4,491	1,855	269	265	140	216	2,310	2,520	702	978	1,711	3,296
Kobe	3,809	1,426	1,389	1,224	603	712	1,657	1,087	1,257	1,290	1,129	1,522
Kyoto	3,717	1,426	1,284	1,262	708	745	2,398	1,704	1,346	1,261	1,114	1,484
Fukuoka	1,297	475	2,250	2,045	989	1,178	1,496	1,202	1,699	1,841	814	1,007
Hiroshima	2,038	867	2,032	1,811	1,033	1,173	1,619	1,188	1,243	1,367	1,311	1,655
Sendai	6,671	2,658	701	640	408	511	2,441	1,932	665	785	2,492	4,183

	Buri		Ika		Tako		Ebi		Asari		Hotate	
	Expense	Quantity	Expense	Quantity	Expense	Quantity	Expense	Quantity	Expense	Quantity	Expense	Quantity
National	3,249	2,038	2,681	2,866	1,346	790	3,527	2,028	1,079	1,118	1,558	973
Tokyo	2,857	1,697	2,273	2,215	1,405	745	3,187	1,712	1,433	1,368	1,776	914
Yokohama	3,369	2,180	2,281	2,519	1,273	674	3,753	1,817	1,296	1,202	1,937	983
Osaka	3,530	2,070	2,881	2,697	1,903	1,169	4,127	2,366	984	931	1,480	751
Nagoya	3,350	1,976	2,598	2,835	1,507	823	4,401	2,447	1,664	1,615	1,728	868
Sapporo	936	683	2,822	4,373	1,146	969	3,284	2,055	658	642	2,775	2,203
Kobe	3,965	2,148	2,494	2,225	2,203	1,289	4,206	2,116	840	722	1,359	682
Kyoto	4,126	2,466	3,154	2,801	1,883	998	4,517	2,491	993	972	1,442	792
Fukuoka	3,153	1,984	2,283	1,835	1,013	516	3,195	1,803	1,022	1,228	927	513
Hiroshima	4,358	2,320	2,733	1,970	1,612	883	3,637	2,380	1,139	1,227	1,095	643
Sendai	2,206	1,345	2,391	3,073	1,324	790	2,848	1,554	967	928	1,799	1,261

Note: The expense and the quantity are in Japanese yen and grams, respectively.

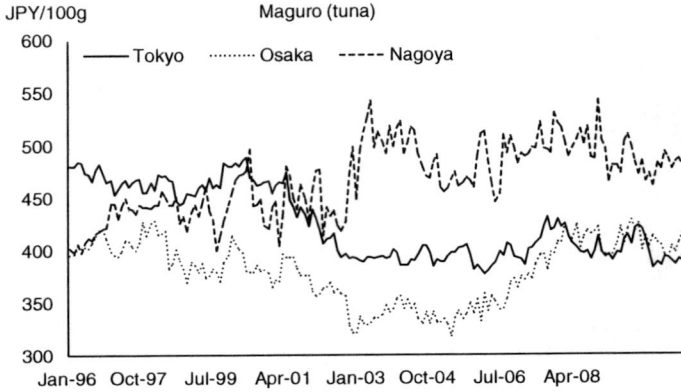

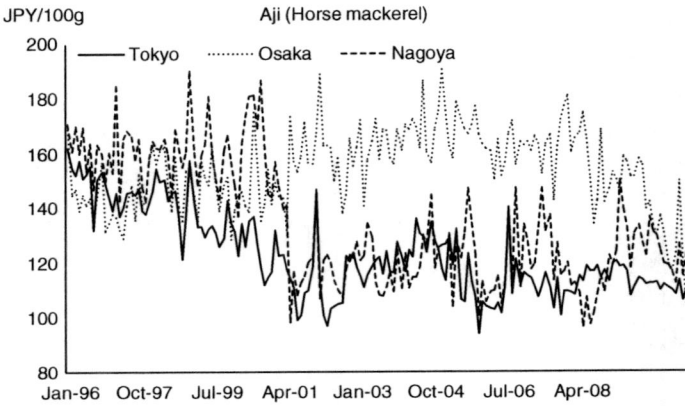

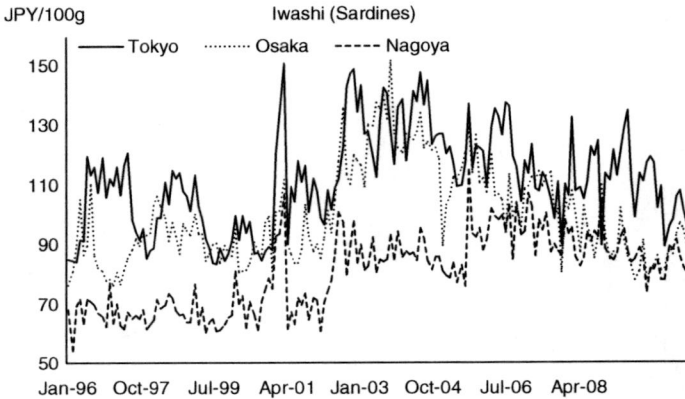

Figure 12.1 Fish retail price (January 1996–December 2009).

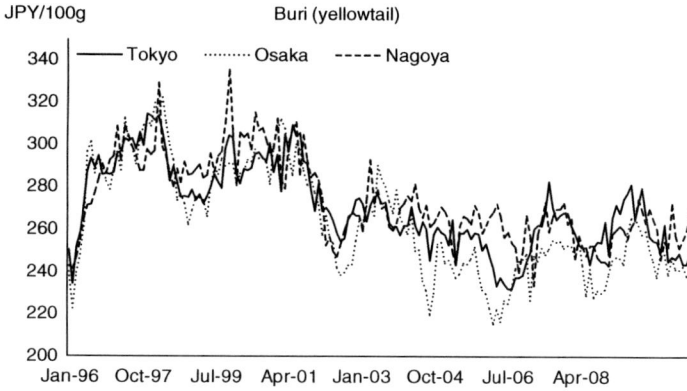

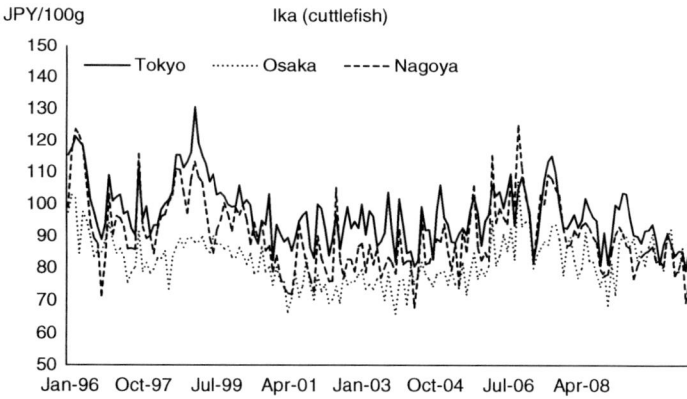

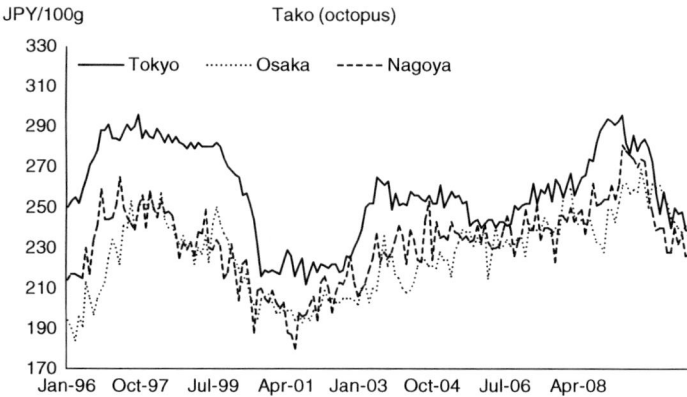

Figure 12.1 Continued.

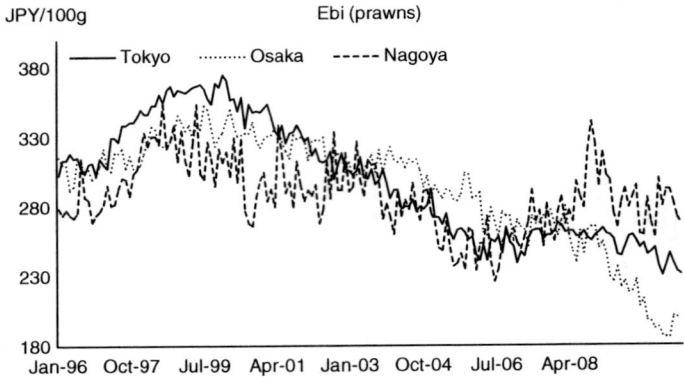

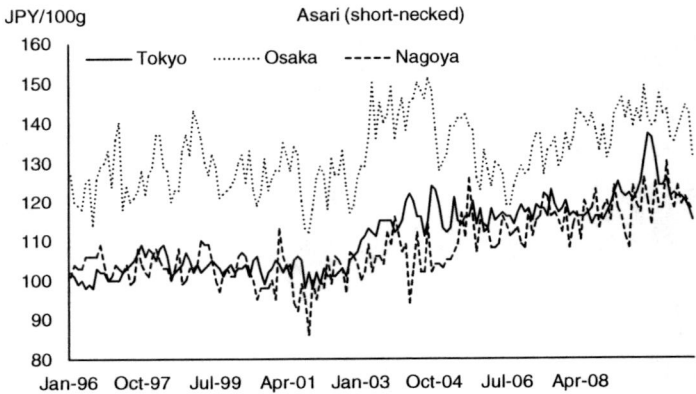

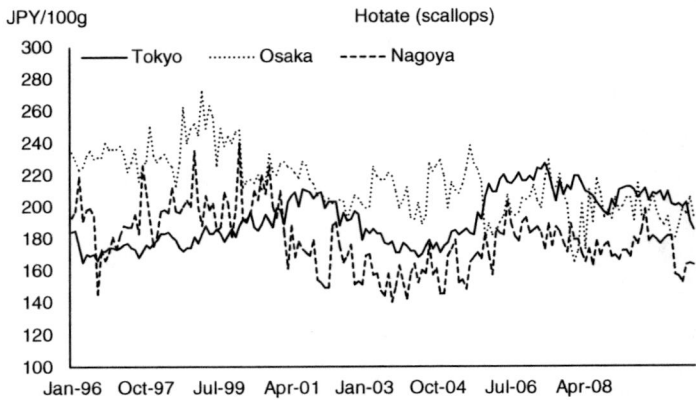

Figure 12.1 Continued.

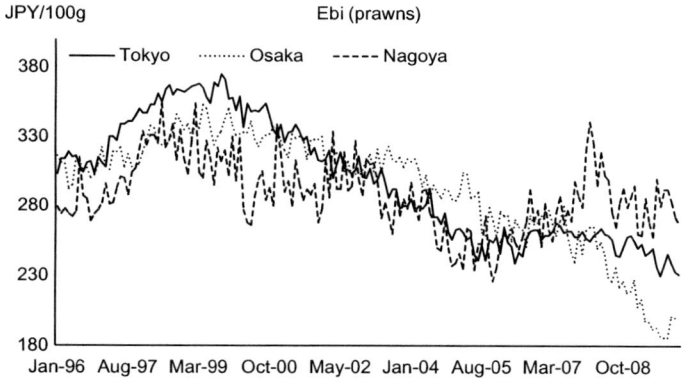

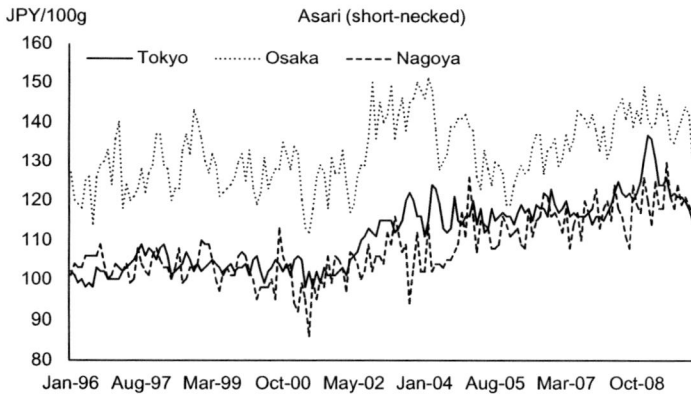

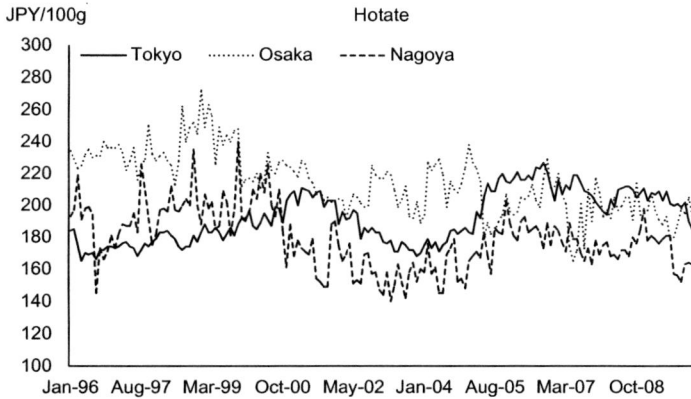

Figure 12.1 Continued.

Methods

We mainly follow the method used by Plourde and Watkins (1998) for comparing the price volatilities among the crude oil price and other commodity prices, such as copper, gold, and wheat. We first calculate the descriptive statistics of the monthly rates of price change for all fish prices in the study and we identify the fish that have the largest and smallest variances for their price change. We do this for ten cities, as explained in the previous section, and compare the variance and median where the fish prices had the largest and smallest variances in their monthly rates of price change.

The monthly rate of price change is calculated by taking the natural log of the fish price and taking the log price difference with the price of the previous month: price change $= \ln P(t) - \ln P(t - 1)$. For every city examined in the study we conducted mainly two types of statistical tests on the fish prices whose variance of the monthly rates of price change was the largest and the smallest among the 12 fish species. One is the dispersion test, which will statistically identify whether the price volatility of the fish with the largest variance for its price change is higher than the price volatility of the fish with the smallest variance for its price change. The other test is the location test. In this test we will find out whether the median of the price change is different between the fish price whose variances of the monthly rates of price change was the largest and the smallest.

We used three types of tests for the dispersion test: the Levene (1960) test, the Bartlett (1937) test, and the Brown and Forsythe (1974) test. All of these tests are used to assess whether the variances in different data samples have the same variance. The Levene test is more powerful when the data are non-normal, and the Bartlett test is more useful if the data are normal. The Brown-Forsythe test is a modified version of the Levene test, and it tests the null hypothesis of equal variance using the analysis of variance based on the medians of the group sample instead of the means. We use the Mann–Whitney–Wilcoxon (MWW) test for the location test. The MWW test assesses whether the underlining distributions of the group samples are the same. We test whether the central tendencies between the populations of the changes in fish prices are the same by testing the null hypothesis that the distributions of the changes in the fish prices for different fish species are equal. Accepting the null hypothesis implies that the median of the changes in the fish prices are the same between different fish species.

Table 12.2 depicts the descriptive statistics of the monthly rates of price change for all fish species used in the study. The table reveals that in most of the ten cities the variances for the monthly rates of price change are the largest for *iwashi* and *sanma*, while those for *maguro* and *buri* are the smallest.

Based on these descriptive statistics, Table 12.3 summarizes the fish price whose variances for the monthly rates of price change were the largest and the smallest among the fish prices examined. Using these fish species with the maximum and minimum price variances among the 12 fish prices, we conducted the (above-mentioned) dispersion and location tests to see if the price variance and median of the price changes between these fish species are the same.

Table 12.2 Descriptive statistics: monthly rates of price change

Species	Mean	Median	Variance	Skewness	Kurtosis
Tokyo					
Maguro	−0.00127	0.00000	0.00033	−0.12179	0.28428
Aji	−0.00232	−0.00516	0.00512	−0.15763	2.33116
Iwashi	0.00074	0.00046	0.01099	−0.62228	3.43826
Karei	−0.00223	0.00110	0.00396	0.07232	0.79031
Saba	−0.00075	0.00144	0.00338	−0.00414	0.67794
Sanma	−0.00166	−0.00627	0.00778	0.05123	3.71436
Buri	−0.00001	−0.00009	0.00088	0.02648	0.46530
Ika	−0.00168	−0.00151	0.00577	0.19404	0.36615
Tako	−0.00027	0.00000	0.00051	−0.50557	0.35002
Ebi	−0.00162	0.00000	0.00068	−0.06038	0.30407
Asari	0.00072	0.00000	0.00063	0.01201	0.19359
Hotate	0.00003	0.00000	0.00080	0.00350	0.35895
Yokohama					
Maguro	0.00061	0.00000	0.00044	1.27396	5.82957
Aji	−0.00244	−0.00929	0.00966	0.01233	1.11470
Iwashi	0.00161	0.00759	0.01597	0.13599	0.73359
Karei	−0.00154	0.00061	0.00509	−0.15743	0.07828
Saba	0.00186	0.00392	0.01178	0.43492	1.49128
Sanma	−0.00147	0.00004	0.01160	−0.13629	3.37646
Buri	0.00043	0.00171	0.00179	0.06924	0.50778
Ika	−0.00095	−0.00104	0.00836	−0.00701	2.10645
Tako	0.00096	0.00000	0.00097	0.85273	4.90429
Ebi	−0.00140	−0.00348	0.00149	0.33747	1.12199
Asari	0.00263	0.00000	0.00130	−0.07257	1.16483
Hotate	0.00160	0.00000	0.00180	−0.09390	0.77212
Osaka					
Maguro	−0.00127	0.00000	0.00075	−0.21935	0.43490
Aji	−0.00347	−0.00464	0.00688	0.10265	0.51299
Iwashi	0.00074	0.00297	0.00852	−0.20204	0.63622
Karei	−0.00223	−0.00430	0.00411	0.63899	1.53175
Saba	−0.00075	−0.00657	0.00712	−0.20699	1.89806
Sanma	−0.00166	0.00643	0.02234	−0.30093	2.95545
Buri	−0.00001	−0.00249	0.00119	0.48536	0.94793
Ika	−0.00168	0.00418	0.00601	0.17212	0.26766
Tako	−0.00027	0.00000	0.00111	0.31288	0.88811
Ebi	−0.00162	0.00000	0.00121	−0.01603	0.28537
Asari	0.00072	0.00717	0.00212	−0.40119	0.37111
Hotate	0.00003	0.00000	0.00328	0.10583	1.57978
Nagoya					
Maguro	0.00105	0.00225	0.00165	−0.18051	1.37551
Aji	−0.00194	0.00148	0.01064	−0.38422	0.87012
Iwashi	0.00097	0.00229	0.01148	−0.48673	4.37267
Karei	−0.00175	−0.00349	0.00192	−0.04333	0.29209
Saba	0.00158	0.00550	0.00691	0.18837	0.95880
Sanma	−0.00162	−0.00097	0.02009	1.15946	13.04077
Buri	0.00061	0.00374	0.00138	−0.12676	0.65984

Continued

Table 12.2 Continued

Species	Mean	Median	Variance	Skewness	Kurtosis
Ika	-0.00153	-0.01122	0.01000	0.55320	1.20858
Tako	0.00035	0.00000	0.00179	0.02215	0.04754
Ebi	-0.00026	-0.00385	0.00456	0.16794	-0.29871
Asari	0.00088	0.00000	0.00277	0.41328	0.59884
Hotate	-0.00101	0.00000	0.00645	-0.06732	1.02882
Sapporo					
Maguro	-0.00049	0.00000	0.00357	-0.01600	0.60058
Aji	-0.00228	-0.00292	0.00822	-0.16803	0.12051
Iwashi	-0.00009	-0.01111	0.02144	0.83255	3.03126
Karei	-0.00521	-0.00316	0.00651	-0.20633	-0.18518
Saba	-0.00145	0.00592	0.01113	-0.09299	2.15783
Sanma	-0.00095	0.00398	0.01197	-1.02233	4.54585
Buri	-0.00190	-0.00153	0.00207	-0.26274	3.04523
Ika	0.00193	-0.00087	0.01676	0.15099	3.66020
Tako	0.00026	0.00000	0.00354	0.61039	3.15472
Ebi	-0.00105	0.00000	0.00330	-0.03610	0.41960
Asari	-0.00022	0.00000	0.00678	-0.39054	1.23918
Hotate	-0.00096	0.00000	0.00819	0.04925	-0.45035
Kobe					
Maguro	0.00124	0.00000	0.00074	0.01198	-0.06912
Aji	0.00207	-0.00381	0.01524	0.38599	0.97673
Iwashi	-0.00081	0.00124	0.01120	0.40478	2.80936
Karei	-0.00047	-0.00803	0.00258	0.49789	2.28846
Saba	0.00129	-0.00044	0.00773	0.18043	1.32305
Sanma	-0.00051	0.00627	0.02235	0.12192	1.78614
Buri	-0.00020	-0.00079	0.00146	0.16184	0.57421
Ika	-0.00101	-0.00005	0.00595	0.07742	1.08820
Tako	0.00025	0.00000	0.00239	-0.00514	1.05352
Ebi	-0.00191	0.00000	0.00105	-0.15343	1.57464
Asari	0.00105	0.00000	0.00253	0.23766	0.55189
Hotate	-0.00226	0.00000	0.00478	0.10850	1.14313

Results and conclusions

The results for the dispersion and location tests conducted between the fish prices whose variances for the monthly rates of price change were the largest and smallest are presented in Table 12.4. According to the three dispersion tests (Levene, Bartlett, and Brown-Forsythe tests), in all cities the fish price with the largest price variance is more dispersed than is the price with the smallest price variance. This indicates that in all ten cities the fish prices whose variances are

Table 12.3 Maximum and minimum price variance

	Fish price with maximum variance		Fish price with minimum variance	
	Species	*Variance*	*Species*	*Variance*
Tokyo	*Iwashi*	0.01099	*Maguro*	0.00033
Yokohama	*Iwashi*	0.01597	*Maguro*	0.00044
Osaka	*Sanma*	0.02234	*Maguro*	0.00075
Nagoya	*Sanma*	0.02009	*Buri*	0.00165
Sapporo	*Iwashi*	0.02144	*Buri*	0.00330
Kobe	*Sanma*	0.02235	*Maguro*	0.00074
Kyoto	*Sanma*	0.01786	*Maguro*	0.00146
Fukuoka	*Aji*	0.01297	*Maguro*	0.00150
Hiroshima	*Ika*	0.02740	*Buri*	0.00163
Sendai	*Iwashi*	0.02346	*Asari*	0.00263

relatively large, such as *iwashi* and *sanma*, have higher price volatilities than the fish prices with a relatively small variance such as *maguro* and *buri*.[3]

Table 12.4 also shows the results of the location test conducted between the fish prices with the largest and smallest variances. The results of the MWW tests suggest that in all cities the rates of price change for the fish price with the largest variance and the smallest variances are distributed similarly. This indicates that the medians of the monthly rates of price change between the fish prices with the largest and smallest variances are the same on average.

These results suggest that among the 12 fish prices there are differences in their price volatilities; and price risks for some fish are relatively higher than the others. Among them, the price of *iwashi* and *sanma* had higher monthly rates of price change, while the price of *maguro* and *buri* had lower rates of price change. This implies that when evaluating the price risk from the Japanese fish retail price, the price risk is higher for the *iwashi* and *sanma* fisheries than for the *maguro* and *buri* fisheries. This might imply that fish producers involved in the *iwashi* and *sanma* fisheries are facing markets with higher price risks, and that it is difficult for these producers to stabilize their profits from these fisheries compared to those engaged in the *maguro* and *buri* fisheries. Hence, it might be preferable for them to balance their catch portfolios and start catching more *maguro* and *buri* to lower their price risks.

It is notable that the prices for *iwashi* and *sanma* were seasonally adjusted, while prices for *maguro* and *buri* were not; however, the former fish prices showed higher price volatilities than the latter fish prices. This indicates that seasonality is an important factor for fish price volatilities to become high, so that considering seasonality might be essential when comparing price risks among fish species. As for the political implications, we need to recognize that some fisheries are involved in this price risk related to seasonality and must be managed in a shorter period, for example annually, and must be treated under special policies.

Table 12.4 Dispersion and location tests

City	Test variable	Levene		Bartlett		Brown-Forsythe		Wilcoxon (MWW)	
		F	Pr > F	Chi-sq	Pr > Chi-sq	F	Pr > F	Z	Pr > Z (one-sided)
Tokyo	*Iwashi* vs *Maguro*	29.5	<0.001	361.6	<0.001	135.3	<0.001	0.2	0.413
Yokohama	*Iwashi* vs *Maguro*	58.5	<0.001	372.6	<0.001	165.2	<0.001	0.6	0.272
Osaka	*Sanma* vs *Maguro*	32.1	<0.001	343.9	<0.001	87.8	<0.001	0.3	0.385
Nagoya	*Sanma* vs *Buri*	2.4	0.124	37.6	<0.001	18.5	<0.001	−0.4	0.342
Sapporo	*Iwashi* vs *Buri*	3.7	0.057	25.7	<0.001	37.1	<0.001	−0.4	0.334
Kobe	*Sanma* vs *Maguro*	42.0	<0.001	346.0	<0.001	114.7	<0.001	0.5	0.302
Kyoto	*Sanma* vs *Maguro*	11.6	0.001	207.1	<0.001	62.5	<0.001	0.4	0.274
Fukuoka	*Aji* vs *Maguro*	38.9	<0.001	175.8	<0.001	88.6	<0.001	−0.7	0.256
Hiroshima	*Ika* vs *Buri*	39.1	<0.001	256.5	<0.001	96.6	<0.001	0.0	0.482
Sendai	*Iwashi* vs *Asari*	51.8	<0.001	168.2	<0.001	109.2	<0.001	0.0	0.491

Another interesting result is that species with the largest variances were all subject to the Total Allowable Catch (TAC) set by the Japanese government. Japan's Ministry of Agriculture, Forestry, and Fisheries (MAFF) currently sets the TAC for the following seven fish species: *aji*, *saba*, *iwashi*, *sanma*, *suketoudara (pollock)*, *snow crab*, and *ika* (MAFF 2010b). The Japanese TAC sets the upper limits for the annual total catches for these fish species caught in the Japanese EEZ. The TAC is essentially imposed on fish species with high commercial value and harvest levels, those that need urgent conservation measures, and species that are potentially targeted by the fish producers of other nations (MAFF 2010b). As seen in Table 12.3, all fish species that had the largest variances (*iwashi*, *sanma*, *aji*, and *ika*) among the 12 fish species are limited in their catch under the TAC. It could be that the current risky situations of fisheries for these fish species are reflected in the Japanese retail markets. Hence, the results of this study may be telling us that it is already difficult for some fish species in the Japanese EEZ to maintain their stocks in a sustainable manner. It could be that this problem of sustaining the fish stocks for this marine area is affecting the retail prices for some of the fish species, making them fluctuate more severely.

Notes

1 The east and west regions are also categorized as north and south regions (see Wessells and Wilen 1994).
2 The prices for *aji*, *iwashi*, *karei*, *saba*, *sanma*, *buri*, and *ika* in the figures are seasonally adjusted using X-12-ARIMA Seasonal Adjustment Program produced by the US Census Bureau.
3 As explained in the second section, the price of *maguro* (tuna) is the price for either yellowfin or bigeyed tuna and does not include the price for bluefin tuna which often faces high price volatilities. The reason why the *maguro* price did not show a high price volatility compared with the other fish prices in this study is perhaps related to this way of specification of the *maguro* price in the retail price survey.

References

Bartlett, M. S., 1937, "Properties of sufficiency and statistical tests," *Proceedings of the Royal Statistical Society Series, A* 160: 268–282.
Brown, M. B. and Forsythe, A. B., 1974, "Robust tests for the equality of variances," *Journal of American Statistical Association*, 69: 364–367.
Food and Agricultural Organization of the United Nations (FAO), 2010, *The State of World Fisheries and Aquaculture 2010*, Rome: FAO, p. 35.
Levene, H., 1960, "Robust tests for equality of variances," in I., Ohlin, S. G., Ghurye, W., Hoeffding, W. G., Madow and H. B., Mann (eds), *Contributions to Probability and Statistics – Essays in Honor of Harold Hotelling*, Stanford: Stanford University Press, pp. 278–292.
Ministry of Agriculture, Forestry, and Fisheries (MAFF), 2010a, *Suisan Hakusho* (White Paper on Fisheries), Tokyo: Agriculture and Forestry Statistics Publishing Inc., pp. 10–18.

——, 2010b, *Kaiyousei-seibutu no hozon oyobi kanri ni kansuru kihonkeikaku* (Basic Plan for Conservation and Management of Marine Living Resources), http://www.jfa.maff.go.jp/j/suisin/s_tac/pdf/kihon_keikaku0525.pdf (Accessed: 6 February 2012).

Plourde, A. and Watkins, G. C., 1998, "Crude oil prices between 1985 and 1994: How volatile in relation to other commodities?" *Resource and Energy Economics*, 20(3): 245–262.

Regnier, E., 2007, "Oil and energy price volatility," *Energy Economics*, 29(3): 405–427.

Statistics Bureau, Director-General for Policy Planning and Statistical Research and Training Institute., 2011, *Nihon no Tokei* (Statistics of Japan), Tokyo: Statistics Bureau, Chapter 2.

Wessells, C. and Wilen, J. E., 1994, "Seasonal patterns and regional preferences in Japanese household demand for seafood," *Canadian Journal of Agricultural Economics*, 42(1): 87–103.

13 Market delineation among the Japanese retail fish markets

Kentaka Aruga and Shunsuke Managi

Introduction

Using market mechanisms to protect marine resources is becoming common among various countries such as Iceland, Australia, and New Zealand. For example, many countries have recently introduced individual transferable quota (ITQ) systems (Arnason 2005), which allow people to trade their fishing rights, thereby creating new markets for fishing rights. In order to conduct such polices using market mechanisms, it is important to know how fish markets are integrated among different regional markets so that policymakers can figure out whether different policies are required for different regions for them to work more effectively.

Yagi and Managi (2011) show the potential merit of ITQ and cost reductions in Japan. Because of mismanagement and the subsidy policy, Japanese fishery is not sustainable. Therefore, it is important to understand what makes the current market efficient. It is known that preferences for fish species among Japanese people are different among different regions, but not many studies have tested – using a time series for retail fish prices – to find out the long-run and short-run linkages among different regional fish markets. Hence, in this chapter we test the price linkages among different regional fish markets in Japan and delineate the Japanese fish market by different regional markets to see which regional markets can be integrated and managed under the same policy. In this chapter we use the retail fish prices of the ten highest-populated cities in Japan for *aji* (horse mackerel), *iwashi* (sardines), and *sanma* (saury). These species are all managed under the total allowable catch (TAC) system in Japan because the Japanese government recognizes the need to conserve these fish species, based on their current stock situation.

All of these fish are consumed nationwide in Japan and can be found in most retail shops in the ten cities investigated in the study. However, the price movements among the ten cities for these fish species might be different if regional markets are based on their local demand preference or the supply system. The Japanese fish markets are different from other commodity markets in the following aspects. First, as sushi is very popular in Japan, freshness is an important factor for fish price (see Carroll *et al.* 2001) and it is more common for local fish retailers to obtain their fish from their local ports. Second, although recent developments in fish-storage technologies have made it easier for retailers to obtain fish from more distant areas, the cost of transportation becomes a barrier to acquire fish from other regions. Finally, as suggested by the study of Wessells and Wilen (1994), regional preferences are different for particular fish species in Japan and such preferences are often related to the historical habits of the regions. Hence, it could be that fish markets among different regions would not share price information, and different regional markets have to be managed differently to implement an effective policy.

However, it is still possible to think that there would be price linkages among some of the regional fish markets. First, markets can share price information even when there are no physical trades among markets, because people involved in the market often refer to the prices of other markets for their price-discovery process. Second, even when there are price differences among different regional markets, they could still have similar price paths if the underlying factors affecting the regional markets are the same. For example, policies like the TAC system could affect nationwide fish prices if, in some circumstances, a certain fishery reached its TAC and no further catch was allowed in this fishery. Not just policies, but weather or temperature change too could also affect the nationwide fish markets, and if fish prices are affected by common factors, regional markets could have similar price movements. Finally, it is likely that a market linkage will exist among neighboring regions where transportation costs are low. As for the case with other commodities, there are arbitrageurs who buy the goods from inexpensive markets and sell them to markets where the same goods are sold at a higher price. This activity of the arbitrageur often brings the markets to move together so that price information is shared in the long run.

There are several studies that have tested the market integration of fish markets (Asche *et al.* 1999; Gordon and Hannesson 1996; Squires *et al.* 1989) but these studies investigate the price linkage among international fish markets, and there are still only a limited number of studies focusing on the linkage among the Japanese regional fish markets. Thus, this study analyses the price relationships among the ten highest-populated cities in Japan and tries to configure how the Japanese regional fish markets can be delineated. This will not only provide a valuable resource for conducting effective policy to conserve the marine resources in Japan, but it will be helpful for understanding more about one of the most active fish markets in the world.

In the next section we explain the methods and data used in the study. The third section presents the results of the analysis. Finally, in the last section, we reveal the conclusions and implications of the study.

Methods and data

For testing the long-run relationship between different regional markets we use the cointegration method developed by Søren Johansen (1991). This method is useful for characterizing the existence of a long-run relationship among different price series, and it has been one of the most popular methods for identifying price linkages among different commodity prices. If cointegration is found among different regional fish prices it would mean that these markets do not deviate in the long run but, rather, follow similar price paths throughout the test period. Hence, having this long-run relationship would imply that these markets share price information and are somewhat integrated.

The cointegration test requires the price series to be integrated of the same order, which means that all variables used in the cointegration test have to become stationary in the same order.[1] We used the augmented Dickey-Fuller (1979) unit root test for this purpose. This tests the null hypothesis of having a unit root against the alternative hypothesis of not having a unit root in the price series. We conducted the cointegration tests only for the price series whose order of integration became the same based on the results of this unit root test. We did not include the price series whose order of integration became different from the other series.

The short-run price relationships among the ten regional markets are identified by the Toda and Yamamoto (1995) modified Granger causality test. Sims *et al.* (1990) suggest that the results of causality tests become spurious when the variables used in the causality test are non-stationary; however, the Toda and Yamamoto method overcomes this problem. According to Toda and Yamamoto (1995), differencing of the price series in order to make the series stationary is not necessary for the estimation of the vector autoregressive (VAR) model in their causality test. In the Toda and Yamamoto procedure, the Granger causality test is tested using $k + d$ lags in the VAR model where k is the statistically optimal lag order in the VAR model and d is the maximal order of integration of the price series in the model. In our study, we used the Akaike Information Criteria (AIC) for identifying the order of k in the VAR model.

The data for the Japanese retail fish prices are obtained from the Japanese retail price survey, which is the same source as for the previous chapter. As explained in the data section of the previous chapter, the data contained 70 cities whose population was more than 150,000 or capitals of the 47 sub-national jurisdictions. Among these 70 cities, we used the retail prices of the cities that are the top ten highest populations in Japan (Statistics Bureau 2011): Tokyo, Yokohama, Osaka, Nagoya, Sapporo, Kobe, Kyoto, Fukuoka, Hiroshima, and Sendai. We use the seasonally adjusted price data on *aji*, *iwashi*, and *sanma* prices in these ten cities. The seasonal adjustments of the price series are conducted by applying the X-12 ARIMA Seasonal Adjustment Program developed by the US Census Bureau.

Among the ten cities listed above, Tokyo and Yokohama belong to the so-called *Kantou* region, while Osaka, Kobe, and Kyoto belong to the *Kinki* region. If geographic location is an important factor for determining regional retail fish prices, it is likely that a price relationship will be found between the cities that belong to these regions. On the other hand, markets of cities that are distanced from other cities, such as Sapporo (locates in the far north) and Fukuoka (locates in the far south), may not have price relationships with other markets.

Table 13.1 Unit root tests

	Aji		Iwashi		Sanma	
	Level	First difference	Level	First difference	Level	First difference
Tokyo	−2.00	−3.50*	−3.07*	−5.10*	−1.15	−6.34*
Yokohama	−1.66	−4.78*	−1.68	−4.82*	−1.22	−6.14*
Osaka	−1.37	−12.45*	−1.97	−10.88*	−3.81*	−9.02*
Nagoya	−2.72	−8.90*	−2.24	−4.99*	−2.49	−5.63*
Sapporo	−3.42*	−11.01*	−6.75*	−10.39*	−0.77	−6.10*
Kobe	0.20	−9.68*	−2.96*	−21.65*	−1.46	−7.14*
Kyoto	−2.70	−3.87*	−2.66	−9.52*	−3.04*	−6.61*
Fukuoka	−1.55	−4.81*	−4.48*	−7.53*	−1.26	−6.06*
Hiroshima	−1.84	−8.14*	−2.62	−8.59*	−0.79	−7.00*
Sendai	−9.27*	−5.79*	−2.64	−10.53*	−3.16*	−6.16*

Note: * denotes significance at the 5% level.

Results

Initially, the ADF unit root tests are applied to all price series used in the study. The ADF tests are conducted both on the levels and first differences of the price series. The results of these tests are presented in Table 13.1. For *aji*, the fish prices of Sapporo and Sendai were stationary for both levels and first differences. Similarly, for *iwashi* the prices of Tokyo, Sapporo, Kyoto, and Fukuoka and, for *sanma*, the prices of Osaka, Kobe, and Sendai were stationary even for the levels of the price series. On the other hand, all other price series were not rejected in the ADF test of having a unit root for the levels of the price series, while they became stationary after differencing. Because cointegration requires the price series to be stationary in the same order, we performed the cointegration test on the price series that were integrated of order 1 based on the results in Table 13.1.

Tables 13.2 through 13.4 show the results of the cointegration and causality tests conducted for *aji*, *iwashi*, and *sanma* among the Japanese regional fish markets. For the *aji* market, cointegration relationships were found between Tokyo and Kobe; Yokohama and Nagoya; Yokohama and Kobe; and Kobe and Hiroshima (see Table 13.2).[2] These results indicate that there is a long-run price linkage between these cities. The result of the causality test for *aji* reveals that the fish markets for Nagoya and Kyoto play an important role in transmitting price information to other cities. For example, looking at the results of the causality tests between Yokohama and Nagoya and between Yokohama and Kyoto in Table 13.2, there are short-run unidirectional price information flows from Nagoya to Yokohama and from Kyoto to Yokohama.

For the *iwashi* market, cointegration existed only between Nagoya and Hiroshima (see Table 13.3). It seems that long-run information flows among the regional *iwashi* markets are very limited, which implies that regional markets

Table 13.2 Cointegration and causality tests for *aji*

Variables	H$_0$: rank $= r$	Trace test	Max test	Granger Causality test	Chi-sq test
Tokyo and Yokohama	$r = 0$	15.76*	11.24	Tokyo $\neq\rightarrow$ Yokohama	2.31*
	$r \leq 1$	4.52*	4.52*	Yokohama $\neq\rightarrow$ Tokyo	1.17
Tokyo and Osaka	$r = 0$	6.18	6.06	Tokyo $\neq\rightarrow$ Osaka	0.51
	$r \leq 1$	0.12	0.12	Osaka $\neq\rightarrow$ Tokyo	1.33
Tokyo and Nagoya	$r = 0$	28.72*	21.38*	Tokyo $\neq\rightarrow$ Nagoya	3.94*
	$r \leq 1$	7.34*	7.34*	Nagoya $\neq\rightarrow$ Tokyo	1.38
Tokyo and Kobe	$r = 0$	7.68	7.12	Tokyo $\neq\rightarrow$ Kobe	0.88
	$r \leq 1$	0.56	0.56	Kobe $\neq\rightarrow$ Tokyo	2.56*
Tokyo and Kyoto	$r = 0$	16.42*	13.01	Tokyo $\neq\rightarrow$ Kyoto	0.82
	$r \leq 1$	3.41	3.41	Kyoto $\neq\rightarrow$ Tokyo	1.64
Tokyo and Fukuoka	$r = 0$	12.95	9.10	Tokyo $\neq\rightarrow$ Fukuoka	1.37
	$r \leq 1$	3.85*	3.85*	Fukuoka $\neq\rightarrow$ Tokyo	0.81
Tokyo and Hiroshima	$r = 0$	8.77	7.28	Tokyo $\neq\rightarrow$ Hiroshima	1.63
	$r \leq 1$	1.49	1.49	Hiroshima $\neq\rightarrow$ Tokyo	1.72
Yokohama and Osaka	$r = 0$	9.46	7.56	Yokohama $\neq\rightarrow$ Osaka	0.06
	$r \leq 1$	1.90	1.90	Osaka $\neq\rightarrow$ Yokohama	1.24
Yokohama and Nagoya	$r = 0$	23.63*	20.15*	Yokohama $\neq\rightarrow$ Nagoya	1.45
	$r \leq 1$	3.48	3.48	Nagoya $\neq\rightarrow$ Yokohama	2.52*
Yokohama and Kobe	$r = 0$	11.44	11.42	Yokohama $\neq\rightarrow$ Kobe	1.02
	$r \leq 1$	0.01	0.01	Kobe $\neq\rightarrow$ Yokohama	1.87*
Yokohama and Kyoto	$r = 0$	24.98*	21.22*	Yokohama $\neq\rightarrow$ Kyoto	1.36
	$r \leq 1$	3.77	3.77	Kyoto $\neq\rightarrow$ Yokohama	2.10
Yokohama and Fukuoka	$r = 0$	16.33	14.39	Yokohama $\neq\rightarrow$ Fukuoka	2.00*
	$r \leq 1$	1.94	1.94	Fukuoka $\neq\rightarrow$ Yokohama	2.37*
Yokohama and Hiroshima	$r = 0$	12.54	10.59	Yokohama $\neq\rightarrow$ Hiroshima	1.92
	$r \leq 1$	1.96	1.96	Hiroshima $\neq\rightarrow$ Yokohama	1.18
Osaka and Nagoya	$r = 0$	10.71	8.88	Osaka $\neq\rightarrow$ Nagoya	1.06
	$r \leq 1$	1.84	1.84	Nagoya $\neq\rightarrow$ Osaka	0.42
Osaka and Kobe	$r = 0$	4.19	4.12	Osaka $\neq\rightarrow$ Kobe	0.43
	$r \leq 1$	0.06	0.06	Kobe $\neq\rightarrow$ Osaka	0.15
Osaka and Kyoto	$r = 0$	7.37	6.57	Osaka $\neq\rightarrow$ Kyoto	1.59
	$r \leq 1$	0.80	0.80	Kyoto $\neq\rightarrow$ Osaka	1.54
Osaka and Fukuoka	$r = 0$	7.48	6.26	Osaka $\neq\rightarrow$ Fukuoka	1.30
	$r \leq 1$	1.22	1.22	Fukuoka $\neq\rightarrow$ Osaka	1.65
Osaka and Hiroshima	$r = 0$	6.63	4.89	Osaka $\neq\rightarrow$ Hiroshima	1.92
	$r \leq 1$	1.74	1.74	Hiroshima $\neq\rightarrow$ Osaka	0.81
Nagoya and Kobe	$r = 0$	7.23	7.17	Nagoya $\neq\rightarrow$ Kobe	0.36
	$r \leq 1$	0.06	0.06	Kobe $\neq\rightarrow$ Nagoya	0.94
Nagoya and Kyoto	$r = 0$	38.50*	33.07*	Nagoya $\neq\rightarrow$ Kyoto	4.05*
	$r \leq 1$	5.44*	5.44*	Kyoto $\neq\rightarrow$ Nagoya	1.97

Continued

Table 13.2 Continued

Variables	H_0: rank $= r$	Trace test	Max test	Granger Causality test	Chi-sq test
Nagoya and	$r = 0$	27.22*	22.88*	Nagoya $\neq\!\rightarrow$ Fukuoka	2.36
Fukuoka	$r \leq 1$	4.34*	4.34*	Fukuoka $\neq\!\rightarrow$ Nagoya	3.25*
Nagoya and	$r = 0$	9.31	6.77	Nagoya $\neq\!\rightarrow$ Hiroshima	1.74
Hiroshima	$r \leq 1$	2.54	2.54	Hiroshima $\neq\!\rightarrow$ Nagoya	0.70
Kobe and	$r = 0$	9.84	9.84	Kobe $\neq\!\rightarrow$ Kyoto	1.76
Kyoto	$r \leq 1$	0.00	0.00	Kyoto $\neq\!\rightarrow$ Kobe	0.65
Kobe and	$r = 0$	8.13	8.06	Kobe $\neq\!\rightarrow$ Fukuoka	2.52*
Fukuoka	$r \leq 1$	0.07	0.07	Fukuoka $\neq\!\rightarrow$ Kobe	3.31*
Kobe and	$r = 0$	4.03	4.01	Kobe $\neq\!\rightarrow$ Hiroshima	1.79
Hiroshima	$r \leq 1$	0.02	0.02	Hiroshima $\neq\!\rightarrow$ Kobe	1.30
Kyoto and	$r = 0$	31.02*	25.55*	Kyoto $\neq\!\rightarrow$ Fukuoka	1.82
Fukuoka	$r \leq 1$	5.47*	5.47*	Fukuoka $\neq\!\rightarrow$ Kyoto	3.33*
Kyoto and	$r = 0$	16.74*	14.00	Kyoto $\neq\!\rightarrow$ Hiroshima	1.09
Hiroshima	$r \leq 1$	2.74	2.74	Hiroshima $\neq\!\rightarrow$ Kyoto	2.40*
Fukuoka and	$r = 0$	9.84	7.41	Fukuoka $\neq\!\rightarrow$ Hiroshima	1.86
Hiroshima	$r \leq 1$	2.43	2.43	Hiroshima $\neq\!\rightarrow$ Fukuoka	0.59

Note: * represents significance at the 5% level; $\neq\!\rightarrow$ denotes that the variable does not Granger cause the other.

for *iwashi* move independently in the long run. The causality test also indicated that only there were only a few short-run price linkages among the regional markets. Some causalities are found between Yokohama and Nagoya, Yokohama and Sendai, and Osaka and Kobe but no short-run causalities were found between other cities.

Finally, see Table 13.4 for the results of the cointegration and causality tests for the *sanma* market. Compared to the *aji* and *iwashi* markets, cointegration held between most of the cities for the *sanma* market. This suggests that some regional *sanma* markets can be integrated and that a long-run information flow exists among the regional markets for *sanma*. In Table 13.4, it is noticeable that although cointegration is sustained between Tokyo and Yokohama, this condition did not occur between Tokyo and Kyoto, or between Yokohama and Kyoto. This result may be reflecting the historical differences in the consumers' preferences between the *Kantou* and *Kansai* regions, and so the *sanma* market can be delineated by the geographical segmentation. The causality test also suggested that more short-run information flows existed for the *sanma* market compared to the *aji* and *iwashi* markets.

In summary, more regional markets share price information in both long-run and short-run tests for the *sanma* market compared to the *aji* and *iwashi* markets.

Table 13.3 Cointegration and causality tests for *iwashi*

Variables	H_0: rank $= r$	Trace test	Max test	Granger Causality test	Chi-sq test
Yokohama and Osaka	$r = 0$	19.84*	13.16	Yokohama $\neq\rightarrow$ Osaka	0.37
	$r \leq 1$	6.68*	6.68*	Osaka $\neq\rightarrow$ Yokohama	1.28
Yokohama and Nagoya	$r = 0$	36.70	28.68	Yokohama $\neq\rightarrow$ Nagoya	1.20
	$r \leq 1$	8.02	8.02	Nagoya $\neq\rightarrow$ Yokohama	6.29*
Yokohama and Kyoto	$r = 0$	16.23*	9.90	Yokohama $\neq\rightarrow$ Kyoto	0.58
	$r \leq 1$	6.33*	6.33*	Kyoto $\neq\rightarrow$ Yokohama	0.82
Yokohama and Hiroshima	$r = 0$	10.31	7.30	Yokohama $\neq\rightarrow$ Hiroshima	1.81
	$r \leq 1$	3.01	3.01	Hiroshima $\neq\rightarrow$ Yokohama	0.79
Yokohama and Sendai	$r = 0$	29.88*	20.60*	Yokohama $\neq\rightarrow$ Sendai	1.89
	$r \leq 1$	9.27*	9.27*	Sendai $\neq\rightarrow$ Yokohama	2.94*
Osaka and Nagoya	$r = 0$	14.00	8.01	Osaka $\neq\rightarrow$ Nagoya	0.34
	$r \leq 1$	5.99*	5.99*	Nagoya $\neq\rightarrow$ Osaka	0.82
Osaka and Kyoto	$r = 0$	45.46*	38.83*	Osaka $\neq\rightarrow$ Kyoto	7.87*
	$r \leq 1$	6.63*	6.63*	Kyoto $\neq\rightarrow$ Osaka	0.60
Osaka and Hiroshima	$r = 0$	31.55*	25.27*	Osaka $\neq\rightarrow$ Hiroshima	1.14
	$r \leq 1$	6.28*	6.28*	Hiroshima $\neq\rightarrow$ Osaka	0.39
Osaka and Sendai	$r = 0$	18.45*	12.09	Osaka $\neq\rightarrow$ Sendai	0.63
	$r \leq 1$	6.36*	6.36*	Sendai $\neq\rightarrow$ Osaka	0.46
Nagoya and Kyoto	$r = 0$	10.03	5.57	Nagoya $\neq\rightarrow$ Kyoto	1.89
	$r \leq 1$	4.46*	4.46*	Kyoto $\neq\rightarrow$ Nagoya	1.71
Nagoya and Hiroshima	$r = 0$	20.53*	17.37*	Nagoya $\neq\rightarrow$ Hiroshima	0.20
	$r \leq 1$	3.16	3.16	Hiroshima $\neq\rightarrow$ Nagoya	0.97
Nagoya and Sendai	$r = 0$	26.32*	18.72*	Nagoya $\neq\rightarrow$ Sendai	2.08
	$r \leq 1$	7.60*	7.60*	Sendai $\neq\rightarrow$ Nagoya	0.67
Kyoto and Hiroshima	$r = 0$	40.46*	27.23*	Kyoto $\neq\rightarrow$ Hiroshima	1.94
	$r \leq 1$	13.23*	13.23*	Hiroshima $\neq\rightarrow$ Kyoto	0.39
Kyoto and Sendai	$r = 0$	15.88*	8.73	Kyoto $\neq\rightarrow$ Sendai	0.87
	$r \leq 1$	7.15*	7.15*	Sendai $\neq\rightarrow$ Kyoto	1.03
Hiroshima and Sendai	$r = 0$	33.61*	23.10*	Hiroshima $\neq\rightarrow$ Sendai	0.87
	$r \leq 1$	10.51*	10.51*	Sendai $\neq\rightarrow$ Hiroshima	0.29

Note: * represents significance at the 5% level. $\neq\rightarrow$ denotes that the variable does not Granger cause the other.

The results indicate that some regional markets can be considered integrated for the *sanma* market, while other regional markets have few linkages. It must also be considered that regional factors affecting market prices cannot be ignored in the *aji* and *iwashi* markets. This suggests that the *sanma* market can, to some extent, be managed under the same policy for some regional markets, while market policy for *aji* and *iwashi* must take into account regional factors that affect those markets.

Table 13.4 Cointegration and causality tests for *sanma*

Variables	H_0: rank $= r$	Trace test	Max test	Granger Causality test	Chi-sq test
Tokyo and Yokohama	$r = 0$	28.13*	25.24*	Tokyo $\neq\rightarrow$ Yokohama	1.60
	$r \leq 1$	2.89	2.89	Yokohama $\neq\rightarrow$ Tokyo	1.18
Tokyo and Nagoya	$r = 0$	14.20	11.68	Tokyo $\neq\rightarrow$ Nagoya	1.29
	$r \leq 1$	2.52	2.52	Nagoya $\neq\rightarrow$ Tokyo	1.13
Tokyo and Sapporo	$r = 0$	17.25*	14.43*	Tokyo $\neq\rightarrow$ Sapporo	1.83
	$r \leq 1$	2.82	2.82	Sapporo $\neq\rightarrow$ Tokyo	2.65*
Tokyo and Kobe	$r = 0$	13.51	11.01	Tokyo $\neq\rightarrow$ Kobe	1.40
	$r \leq 1$	2.49	2.49	Kobe $\neq\rightarrow$ Tokyo	3.49*
Tokyo and Fukuoka	$r = 0$	33.14*	29.51*	Tokyo $\neq\rightarrow$ Fukuoka	13.01*
	$r \leq 1$	3.64	3.64	Fukuoka $\neq\rightarrow$ Tokyo	1.66
Tokyo and Hiroshima	$r = 0$	17.26*	16.09*	Tokyo $\neq\rightarrow$ Hiroshima	0.65
	$r \leq 1$	1.17	1.17	Hiroshima $\neq\rightarrow$ Tokyo	6.30*
Yokohama and Nagoya	$r = 0$	22.87*	18.80*	Yokohama $\neq\rightarrow$ Nagoya	0.89
	$r \leq 1$	4.08*	4.08*	Nagoya $\neq\rightarrow$ Yokohama	1.23
Yokohama and Sapporo	$r = 0$	17.19*	14.42*	Yokohama $\neq\rightarrow$ Sapporo	1.06
	$r \leq 1$	2.77	2.77	Sapporo $\neq\rightarrow$ Yokohama	1.95
Yokohama and Kobe	$r = 0$	26.69*	22.84*	Yokohama $\neq\rightarrow$ Kobe	1.26
	$r \leq 1$	3.85*	3.85*	Kobe $\neq\rightarrow$ Yokohama	1.79
Yokohama and Fukuoka	$r = 0$	15.49*	26.09*	Yokohama $\neq\rightarrow$ Fukuoka	12.75*
	$r \leq 1$	3.84	3.83	Fukuoka $\neq\rightarrow$ Yokohama	1.54
Yokohama and Hiroshima	$r = 0$	18.55*	16.57*	Yokohama $\neq\rightarrow$ Hiroshima	1.40
	$r \leq 1$	1.98	1.98	Hiroshima $\neq\rightarrow$ Yokohama	4.77*
Nagoya and Sapporo	$r = 0$	16.09*	13.62	Nagoya $\neq\rightarrow$ Sapporo	1.72
	$r \leq 1$	2.46	2.46	Sapporo $\neq\rightarrow$ Nagoya	1.40
Nagoya and Kobe	$r = 0$	16.46*	13.26	Nagoya $\neq\rightarrow$ Kobe	0.77
	$r \leq 1$	3.20	3.20	Kobe $\neq\rightarrow$ Nagoya	3.43*
Nagoya and Fukuoka	$r = 0$	51.78*	47.91*	Nagoya $\neq\rightarrow$ Fukuoka	18.67*
	$r \leq 1$	3.87*	3.87*	Fukuoka $\neq\rightarrow$ Nagoya	3.34*
Nagoya and Hiroshima	$r = 0$	17.98*	15.97*	Nagoya $\neq\rightarrow$ Hiroshima	3.67*
	$r \leq 1$	2.01	2.01	Hiroshima $\neq\rightarrow$ Nagoya	2.26
Sapporo and Kobe	$r = 0$	18.86	17.43	Sapporo $\neq\rightarrow$ Kobe	1.55
	$r \leq 1$	1.43	1.43	Kobe $\neq\rightarrow$ Sapporo	1.99
Sapporo and Fukuoka	$r = 0$	22.52*	19.26*	Sapporo $\neq\rightarrow$ Fukuoka	10.41*
	$r \leq 1$	3.27	3.27	Fukuoka $\neq\rightarrow$ Sapporo	0.51
Sapporo and Hiroshima	$r = 0$	15.49	11.61	Sapporo $\neq\rightarrow$ Hiroshima	1.91*
	$r \leq 1$	3.84	1.15	Hiroshima $\neq\rightarrow$ Sapporo	2.69*
Kobe and Fukuoka	$r = 0$	36.16*	33.06*	Kobe $\neq\rightarrow$ Fukuoka	10.21*
	$r \leq 1$	3.10	3.10	Fukuoka $\neq\rightarrow$ Kobe	3.00
Kobe and Hiroshima	$r = 0$	23.85*	21.27*	Kobe $\neq\rightarrow$ Hiroshima	3.28*
	$r \leq 1$	2.58	2.58	Hiroshima $\neq\rightarrow$ Kobe	1.27
Fukuoka and Hiroshima	$r = 0$	22.47	19.40	Fukuoka $\neq\rightarrow$ Hiroshima	5.83*
	$r \leq 1$	3.07	3.07	Hiroshima $\neq\rightarrow$ Fukuoka	1.77

Note: * represents significance at the 5% level. $\neq\rightarrow$ denotes that the variable does not Granger cause the other.

A possible reason why a regional linkage persisted more in the *sanma* market compared to *aji* and *iwashi* markets is that *sanma* caught around Japan all falls into a single clade (see Chow *et al.* 2009), the Pacific system, while *aji* and *iwashi* can be categorized into two clades (FRA 2010): the Pacific system and the Tsushima warm-current system.[3] Since 2001, the Japanese government has assessed the status of marine resources in its surrounding marine area for species that have high commercial value and are in need of conservation. Every year, the government publishes the assessment of these marine resources separated by not only species but also by major existing clades for each species. Here, the *sanma* category contains one clade while *aji* and *iwashi* are separated into the above mentioned clades. Hence, it is likely that fishermen involved in the *sanma* fisheries are basically hunting the same *sanma* population, while fishermen in the *aji* and *iwashi* fisheries are capturing their fish from several populations. It could be that differences in the types of *aji* and *iwashi* caught among different regions are large compared to those for *sanma* and this is causing the retail prices of *aji* and *iwashi* to move more independently among different regions.

Conclusions

In this chapter, we tested the long-run and short-run market linkages for the *aji*, *iwashi*, and *sanma* retail markets among the markets of ten highly populated cities in Japan to identify how these fish markets can be delineated. The results provide useful information for policymakers on which to conduct effective policy for conserving marine resources in Japan. The three fish species investigated in this study are already protected under the TAC system, but a more advanced fisheries policy may be required to conserve these resources. Because using market mechanisms is one of the effective ways to implement such marine policies, we identified how regional fish markets are linked for these fish species.

Our test results revealed that only a few market linkages exist among the regional markets for the *aji* and *iwashi* markets, and the regional markets for these fish species move independently both in the long run and short run. This indicates that no nationwide market exists for *aji* and *iwashi* and it is not possible to manage these fish markets under the same price policy or price strategy. These markets depend more on their regional characteristics, and policies affecting their markets must be dealt with separately among different regions. On the other hand, for the *sanma* market, we found that a market linkage persisted more widely among different regions compared to the *aji* and *iwashi* markets, and there also existed some linkages within the geographical region such as the *Kanto* region. This implies that national price policy is easier to implement for the *sanma* market compared to the *aji* and *iwashi* markets, and some policies may be conducted under geographical regions.

In this study, we covered only the fish markets for *aji*, *iwashi*, and *sanma* but in Japan there are many other types of fish that are largely consumed and are economically valuable. Further research is necessary to conduct similar studies for other fish markets to find out how these fish markets are linked among different regions and how they can be delineated as well. As seen in the case of the *aji* and *iwashi* markets, there seem to be regional factors affecting local markets that make them move independently. This tells us that considering the characteristics of differences among different regional markets is important for conducting effective policy to protect the biodiversity of these resources in Japan. Further studies on different fish species will help us to understand whether such regional factors should be considered when conducting market policies to conserve their stocks and would thus provide a useful source for implementing more effective policies. Such studies would also help us to understand how alternative policies such as ITQ are able to make the market efficient and change the fishery to be sustainable.

Notes

1 In this context, "order" means the differencing number performed for the price series to become stationary.
2 These results are based on the trace test, because in general the power of the trace and maximum eigenvalue tests are similar; however, the trace test sometimes performs better than the maximum eigenvalue test (Lutkepohl *et al.* 2001).
3 The word "clade" is often used to express a species or population that contains one ancestor and its descendants. For marine organisms, it stands for species that have common ecological behaviors and sustain the same types of body and race.

References

Arnason, R., 2005, "Property rights in fisheries: Iceland's experience with ITQs", *Reviews in Fish Biology and Fisheries*, 15: 243–264.
Asche, F., Bremnes, H. and Wessells, C. R., 1999, "Product aggregation, market integration, and relationships between prices: an application to world salmon markets", *American Journal of Agricultural Economics*, 81: 568–581.
Carroll, M. T., Anderson, J. L. and Martinez-Garmendia, J., 2001, *Agribusiness*, 17(2): 243–254.
Chow, S., Suzuki, N., Brodeur, R. D. and Ueno, Y., 2009, "Little population structuring and recent evolution of the Pacific saury (Cololabis saira) as indicated by mitochondrial and nuclear DNA sequence data", *Journal of Experimental Marine Biology and Ecology*, 369: 17–21.
Dickey, D. A. and Fuller, W. A., 1979, "Distribution of the estimators for autoregressive time series with a unit root", *Journal of the American Statistical Association*, 74(366): 427–431.
Fisheries Research Agency (FRA), 2010, *Heisei 22 nendo gyoshubetsu keigunbetsu shigen hyouka* (Resource Assessment of fish by type and system for 2010), FRA, http://abchan.job.affrc.go.jp/digests22/index.html (Accessed: 6 February 2012).
Gordon, D. V. and Hannesson, R., 1996, "On prices of fresh and frozen cod fish in European and U.S. markets", *Marine Resource Economics*, 11(4): 223–238.
Johansen, S., 1991, "Estimation and hypothesis testing of cointegration vectors in Gaussian Vector Autoregressive Models", *Econometrica*, 59(6):1551–1580.
Lutkepohl, H., Saikkonen, P. and Trenkler, C., 2001, "Maximum eigenvalue versus trace tests for the cointegrating rank of a VAR process", *Econometrics Journal*, 4(2): 287–310.
Sims, C. A., Stock, J. H., and Watson, M. W., 1990, "Inference in linear time series models with some unit roots", *Econometrica*, 58(1): 113–144.
Squires, D., Herrick, S. F., and Hastie, J., 1989, "Integration of Japanese and United States sablefish markets", *Fishery Bulletin*, 87(2): 341–351.

Statistics Bureau, Director-General for Policy Planning and Statistical Research and Training Institute, 2011, *Nihon no Tokei* (Statistics of Japan), Tokyo: Statistics Bureau, Chapter 2.

Toda, H. and Yamamoto, T., 1995, "Statistical inference in vector autoregressions with possibly integrated processes", *Journal of Econometrics*, 66(1–2): 225–250.

Wessells, C. and Wilen, J. E., 1994, "Seasonal patterns and regional preferences in Japanese household demand for seafood", *Canadian Journal of Agricultural Economics*, 42(1): 87–103.

Yagi, M. and Managi, S., 2011, "Catch limits, capacity utilization and cost reduction in Japanese fishery management", *Agricultural Economic*, DOI: 10.1111/j.1574-0862.2010.00533.x.

Conclusion

Towards biodiversity conservation

Shunsuke Managi and Kei Kabaya

Sustainable use of ecosystem services

Key points mentioned in this book are summarized in this chapter. In essence, biodiversity means the diversity of genes, species, and ecosystems; more precisely, biodiversity is the expression of the diverse "state" of those. Since the Earth Summit in 1992, biodiversity has been attracting global attention to the point where its loss has become one of the most important environmental issues in the world. However, biodiversity conservation is complicated and difficult, because loss of biodiversity is a multi-layered issue caused by various factors at local, national, regional, and global levels, and the countermeasures vary widely from gene preservation to species conservation and protected-area designation.

Studies so far have revealed that more diverse species generate higher productivity and stability of ecosystems, and that species diversity contributes to enhanced resilience in the face of disturbance. Meanwhile, not all species need to be conserved to maintain the function of ecosystems. Nevertheless, it is still unclear which species have a minimum effect on the functions and can be erased.

Conservation of the habitats for specific keystone and umbrella species is required to conserve biodiversity under current financial conditions. In brief, priority in biodiversity conservation should be given to ecosystems, which are the "systems" and "functions," the diversity of which is a part of biodiversity. Thus, the conservation policies focusing on ecosystems will be the keys for efficient and effective biodiversity conservation.

However, ecosystems have been degrading at the most rapid rate ever in the history of the world. Cumulative deforestation and reclamation mean that cultivated land currently dominates one-quarter of the terrestrial surface, and coastal development in the last few decades has caused the destruction of coral reefs and mangroves by 20 percent and 35 percent, respectively. From three to six times more water resources than those of the natural river flows are stored in dam sites globally, and species become extinct at a rate 1,000 times greater than the historical rate. The direct drivers of this ecological degradation include habitat change, climate change, invasive alien species, overexploitation, and pollution. Such drivers are backed by indirect factors, that is, population change, change in economic activities, sociopolitical factors, cultural factors, and technological change.

Increases in population and income will expand absolute resource use, and sociopolitical factors may have impacts on decision making and on education for ecosystem management. Recognizing ecosystems and consumption behaviors may uncover strong links to cultural and religious factors, and technological development can be one of the main drivers of the expansion of cultivated lands and overexploitation. Seeing these indirect factors from different viewpoints, market failures and policy failures are the two main indirect causes.

Services and goods provided by ecosystems are called ecosystem services, which vary widely from food provision to climate regulation, recreation, and soil formation, and their impacts range broadly from local (for example flood regulation) to global (for example air-quality regulation). In essence, various services are provided by ecosystems which are supported by the species.

Degradation of ecosystem services has negative impacts on both developed and developing countries, especially on the poor in the world. Degraded water-purification services due to ecosystem destruction and pollution may increase the possibility of diseases from polluted drinking water and pathogen growth, thereby killing vast numbers of people every year. This situation is made worse by a scarcity of public sanitation. Extinction of plant species due to deforestation and development will mean the loss of opportunities to discover genetic resources for new drug development; and a decrease in fish stock due to overexploitation may lower food production and employment potential in developed countries as well. The disappearance of forests and wetlands will increase risks of climate change, such as the rise in the sea level in coastal areas, higher intensity of natural disasters, and changes in fauna, flora, and crops. The increase in disease in developing countries may spill out to developed countries, and the degradation of ecosystems in the biodiverse tropical forests will devalue ecotourism.

The ecosystem service is a concept which contains not only natural resources directly used but also various regulating services such as flood control and air-quality regulation. Keeping those in mind, global warming can be understood as an issue of sustainable use of climate-regulation services. Likewise, biodiversity conservation can also be recognized as an issue of sustainable use of ecosystem services, with biodiversity itself providing ecosystem services or ecosystem services depending on biodiversity conservation. Clearly framing the issues around us, sustainable development defined in the Brundtland Report – prepared by the World Commission on Environment and Development (WCED), which was founded by the United Nations in 1984 – can be interpreted as sustainable use of ecosystem services; and the (simultaneous) reduction of poverty, defined as lack of opportunities to meet basic needs. As such, the issue of sustainable use of ecosystem services is an extremely important policy issue.

So as to shift from the unsustainable use of ecosystem services, it will be essential to build an economic structure providing greater economic efficiency for sustainable use, rather than an unsustainable one. In doing so, there needs to be a change in economic systems that include the creation of incentives as well as the elimination of subsidies that encourage the degradation of ecosystem services.

In this regard, the Nagoya Protocol and the Aichi Target were concluded at the CBD-COP10 on 30 October 2010. The former contributes to international rule-making on fair and equitable benefit-sharing from the use of genetic resources. The protocol mentions the prior informed consent of the countries providing resources and the implementation of legal measures in each country. The latter is the newly concluded goal for ecosystem conservation from the year 2011 in response to the failure of the 2010 Biodiversity Target, which aims to achieve "by 2010 a significant reduction of the current rate of biodiversity loss at the global, regional and national level." The Aichi Target sets 20 individual goals for each country to achieve, including designation of 17 percent of terrestrial areas and 10 percent of marine areas as protected areas. Additionally, several other decisions were made such as the creation of the Intergovernmental Science and Policy Platform on Biodiversity and Ecosystem Services (IPBES), establishment of a strategy for world plant conservation, and recognition of the importance of rice fields.

The National Strategy for the Conservation and Sustainable Use of Biological Diversity has been formulated four times in Japan. The third National Strategy underlines three crises. The first crisis is species and habitat degradation due to excessive human activities and the second one is degradation of so-called "*satochi-satoyama*" due to insufficient levels of management. The 2010 National Strategy decided by the Cabinet in March 2010 proposes the *SATOYAMA* initiative, which indicated the risks of biodiversity loss caused by degradation of secondary nature – due to insufficient levels of management – and highlighted the importance of *satochi-satoyama*. The third crisis is ecosystem disturbances caused by the introduction of alien species and chemical contamination. Last, the extinction of a number of species and the collapse of ecosystems due to the crisis of global warming are indicated.

Biodiversity and the economic valuation of ecosystem services

What will be required for biodiversity conservation? Substantial funds will be needed, with the burden of expenses imposed on beneficiaries. How is it possible to secure the funds? Currently, economic valuation of ecosystem services attracts attention, and so do the implementation tools, for example payment for ecosystem services (PES hereafter) and biodiversity offset.

PES is actually defined as voluntary trade of clearly defined environmental services or lands providing those services between beneficiaries and service providers, but these days the term PES is used as a collective term to refer to various market-based conservation schemes. What is important is visualizing the invisible values of nature within a market economy. Currently, mechanisms similar to PES are widely implemented for maintaining and improving ecosystem services. Biodiversity offset aims at offsetting the impacts of development works on species and habitats by the creation at neighboring sites of an artificial environment which, to some extent, may have similar functions and quality to the destroyed habitats.

One of the major objectives of economic valuation of biodiversity and ecosystem services is the demonstration of values, thereby integrating them into decision making on business and policies. In this sense, an increase in trade and conservation activities through market mechanisms such as PES will contribute to expanding opportunities for disseminating those values.

In order to achieve this objective, evaluation of the economic value of ecosystem services is imperative. The social costs of the loss of biodiversity need to be understood and the social meaning for biodiversity conservation should be indicated in doing so. Also, frameworks for urging payment of rewards to ecosystem services and sharing benefits from biodiversity have to be built. In this regard, it is vital to evaluate the non-use values of ecosystem services.

However, it will not be easy to evaluate these in monetary terms, as no market prices exist for those ecosystem services. In general, environmental values can be divided into use values, which are benefits from direct and indirect use of the environment, and non-use values, which are generated from leaving the environment as it is: the value of ecosystem services includes both of these. So as to evaluate these, stated preference approaches such as the Contingent Valuation Method (CVM) and choice experiment are required. Economic valuation of ecosystem services contributes to institutional designs in terms of finding problems, comprehending demands, and providing information to decision makers and citizens.

Here, benefit transfer by choice experiments was tested by using the results of surveys conducted in four municipalities. The municipalities have similar types of rice-terrace resources and they participate in the direct payment scheme. As a basis for implementing such a quasi-PES program to conserve rural amenities, benefit valuation and benefit transfer clearly demonstrate the importance of public and private support. Benefit estimates show only a small difference between different policy sites. It continues to be a signal and driving force for the implementation of PES.

The only exception was a function estimated with the data of Kamogawa City. It may be significant that only in Kamogawa City was the survey conducted when rice was actually growing in the rice terraces. The surveys in the three other municipalities were conducted after the rice was reaped. The results of the analyses may indicate that the evaluation reflects the condition of the landscape and the environment at the time of the survey. For example, the evaluation of national land conservation and amenity in Kamogawa City shows a significant difference, while evaluations in the three other municipalities show no significant differences. Evaluation by choice experiments is likely to be influenced by price fluctuations, depending upon when a survey is conducted. Therefore, when conducting benefit transfer, it is necessary to pay special attention to this, perhaps by conducting the surveys at the same time.

Values of recreation and ecotourism derived from biodiversity are also essential. For valuing these services, a spatial econometric approach to the Kuhn–Tucker (KT) model was considered. Our proposed approach is appealing in that it can analyze spatial heterogeneity within the single-structure KT framework, which simultaneously models recreational participation and site selection decisions while allowing for corner solutions and consistency with utility theory. The proposed and standard KT models were applied to a recreation dataset for the national and quasi-national parks in Hokkaido, Japan, and the model results were compared. The empirical analysis showed that both spatial heterogeneity across sites and non-spatial heterogeneity across individuals exist. For welfare analysis, we considered three scenarios. Estimated results show that decreasing the protected species has a larger welfare loss than the one of the site closure of the Shiretoko world heritage site. The results suggest that it is important to account for the value of biodiversity in park management.

We suggest two additional paths of investigation to the spatial approach to the KT model. First, other spatial econometric approaches to the KT model should be investigated. In the spatial econometrics literature, some models have been developed for spatial data, such as the spatial lag model and the spatial autoregressive model. The application of these models to the KT framework may be meaningful. However, it might be complex, because the utility function used in the KT model is not linear with respect to the error term. Second, multi-destination trips should be analyzed. Visitors taking multi-destination trips may prefer neighboring sites to remote sites. Thus, multi-destination trips may cause spatial behavior in recreation demand. We used a simple approach to the multi-destination trips. Nevertheless, a sophisticated analysis of the multi-destination trips, particularly for the spatial KT model, is an important issue for future research.

Biodiversity, which supports ecosystem services, is also known to have positive impacts on net primary production by promoting the efficient use of resources, for example sunlight and nutrition, but there are few studies on the economic value of biodiversity itself and the contribution to GDP thereof. Considering that biodiversity is of global importance, which may affect not only agriculture but also genetic resources and ecotourism, further analysis from a wider perspective such as national GDP will be required.

Econometric analysis on biodiversity, biomass, and productivity models revealed that biodiversity which was estimated from forest areas, rainfall, and temperature would increase biomass in collaboration with inland water areas, which in turn would have positive and significant impacts on the GDP growth rate. Also, scenario analysis demonstrated that unit forest area as well as species could have the highest values in Asia, while unstable climate conditions defined as larger fluctuations of rainfall and temperature could have the biggest impacts in the Middle East.

Economic valuation from the production or supply sides may complement that from the consumption and demand sides, and may contribute to a detailed and accurate economic valuation. In this sense, economic values based on both supply and demand will need to be applied when those are utilized in future "green accounting" schemes.

From a different perspective, determinants of happiness were investigated in relation to environmental degradation. We expect environmental degradation and attachment to nature to affect the level of happiness, and our results imply that environmental degradation above a certain threshold can affect people's level of happiness, and attachment to nature tends to be positively related to happiness. We therefore conclude that both environmental conservation and attachment to nature can increase the level of people's happiness.

Our results suggest that if environmental education can improve people's attachment to nature, it would lead to increased happiness. In addition, although this is beyond the scope of this study, we can confidently say that if people who grow up close to nature tend to have a strong attachment for nature, living surrounded by nature will improve our happiness.

Institutional design of PES

There have been PES-like ideas and frameworks in Japan which are also applied in the modern society with slight modification. The people of Mizuno village in the Kubiki district, Echigo province (today's Niigata prefecture), submitted a request for approval in 1784 to start charcoal production in the community forest, but they were met by strong protests from the 24 villages situated further downstream. The downstream villages were concerned that "forest exploitation would lead to water shortages due to accelerated thaw in spring, and also create a higher risk of sediment discharge in cases of rain." An agreement was reached for Mizuno village to cancel the forest-clearing and charcoal-production plans, by receiving in exchange "50 Ryo (past currency) as a first payment, and a yearly compensation of 150 kilograms of rice." Furthermore, it was stipulated in their agreement that "when forest overgrowth causes deer and boars to ravage the adjacent farmland, tree felling will be conducted under the witness of both upstream and downstream villages, following the upstream farmlands' maintenance needs but respecting the limit of potential impacts to water provision." This case can be seen as a payment by the downstream communities to the upstream communities for the preservation of a stable water supply as a forest ecosystem service which is needed for rice cultivation.

Increased attention to environmental problems as well as the enactment of the government's Decentralization Law in 2000 boosted the introduction of Forest Environment Taxes among local governments. The main aim of this prefectural tax is to secure funds for forest maintenance through wide burden-sharing among citizens who benefit from the water storage and other environmental functions of forests. The first to implement this system was Kochi prefecture which established a prefectural ordinance to introduce the Forest Environment Tax in 2003, followed by Okayama prefecture, Tottori prefecture, and many others. As of April 2009, there were 30 prefectural governments and one city government which had introduced this tax. Likewise, private companies have been involved in water-source forest conservation. In Japan, there have traditionally been many cases of forest maintenance and conservation funded by companies relying on forests for their water supply, such as electronic companies and beer breweries.

Impact assessment of REDD-plus policy

CO_2 emissions from land conversion such as forest loss and degradation are estimated to make up approximately 20 percent of total human induced emissions, which is the second-highest, following emissions from fossil-fuel use. According to TEEB, halving forest loss by 2030 would enable CO_2 emissions reduction by 1.5 to 2.7 Gt annually, and could prevent economic losses equivalent to more than 3.7 trillion US dollars (net present value) due to climate change.

Despite such political importance, there are not many existing studies analyzing the sustainable use of ecosystem services in a quantitative manner, especially those conducting a quantitative impact assessment of policy aiming at sustainable use of ecosystem services, based on a general equilibrium model, which can take account of interactions among various economic actors such as households and firms. With this background, an impact assessment of sustainable forest-use policy, based on a policy-impact assessment model combining a Computable General Equilibrium (CGE) model and a forest stock model reflecting natural forest growth is performed.

In doing so, the business-as-usual (BAU) scenario reflecting the current forest-use situations and the sustainable forest-use (SFU) scenarios aiming at forest-stock conservation are formulated, and an assessment is made of the economic impacts of SFU by comparing the simulation results of the SFU and the BAU scenarios using a multi-sectoral Ramsey-type dynamic CGE model with forestry and lumber sectors. This also covers policy instruments to transform ecosystem services into economic benefits, such as the reduced emissions from deforestation and degradation (REDD), which draws wide attention in the international negotiation process regarding climate change and the PES.

As a result of the assessment, the SFU policy reduces social welfare levels in terms of equivalent variations in the base case with setting the REDD credit price at 4 US dollars per t-CO_2 according to the existing studies. This negative impact is mainly caused by reduced production in the forestry sector, due to restrictions on logging volume, and the wood-product sector that inputs forestry products as the main intermediate input. In this analysis, the capital stock and its underlying households' assets are endogenously determined, and the SFU scenario results in fewer household assets than the BAU scenario does. This study also deals with the issue of appropriate pricing of the REDD credit and PES, revealing that the REDD credit price of 35.6 US dollars per t-CO_2 or PES price for forest ecosystem services other than carbon fixation service 8,500 US dollars per ha, which is equivalent to 3.59 billion US dollars per year of total PES revenues, need to be set so as not to reduce the social welfare level.

REDD-plus, that aims to provide incentives to developing countries to protect and enhance their forest carbon stocks, is now a "hot topic" on the international climate change agenda, and investment for readiness and demonstration activities are taking place in over 30 developing countries. One factor behind the strong interest in REDD-plus is speculation for gains by both developed and developing countries. Developing countries are hoping to receive new investment for forest management from multilateral and bilateral sources. They expect that REDD-plus will provide co-benefits, such as proper national forest inventories, which are yet to be conducted in some countries, and better forest monitoring. Developed countries, on the other hand, are expecting to reduce the costs of achieving mitigation targets by having access to a new cheap source of emissions offsets. In spite of difficult methodological and political issues that still need to be resolved and agreed, the strong interest from developing and developed countries, and the fact that REDD-plus is already relatively well-advanced in the negotiations, suggest that REDD-plus will be realized as part of the future global climate framework.

Biodiversity offset banking mechanism

Among the many types of environmental policies and systems, "tradable allowance schemes" are considered to be effective; under these systems, the rights to utilize the environment or resources are transacted in the market. One typical scheme is the "cap-and-trade scheme," which has been considered to reduce the environmental burden efficiently. In fact, many theoretical and empirical studies have proven the effectiveness of tradable allowance schemes. For example, emission-permits trading schemes have been spreading for the past two decades. The United States first introduced this type of scheme for sulfur-dioxide (SO_2) in 1990, and it succeeded in reducing SO_2 emissions by half.

For the conservation of biodiversity, a kind of tradable-allowance scheme has existed since the 1980s. For example, wetland-mitigation banking schemes (hereafter, "offset scheme") were implemented in the United States: under this scheme, when a "developer" develops a unit of wetland (a unit of land in which indigenous biodiversity is nurtured), it has to restore and (or) create a unit of wetland that is equivalent to the wetland developed. As a result, the quality of wetlands and indigenous biodiversity is preserved in a certain area.

Impacts of the biodiversity offset may include not only habitats and ecosystem functions but also values of human use and cultural value. These impacts should be offset through this mechanism, and no net loss at least, and a net gain at most, should be achieved (Ota *et al.* 2011). However, the value of biodiversity varies across regions due to factors such as the variety of indigenous animals and plants across regions. Even if those animals and plants are the same, environmental values may be different among regions because external benefits can differ. For example, a unit of wetland not only nurtures indigenous plants, but it also influences the environment of adjacent areas. In offsetting the impacts of development, developers do not necessarily have restoration expertise. When this is the case, self-mitigation is very costly, and (or) there is a risk that developers might fail to restore wetlands. The other way is that other players, who are called "bankers," perform the restoration. When a banker restores a unit of wetland, it obtains a unit of "credit" from the authority. Then, if a developer purchases a credit from a banker, the developer is regarded as having restored a unit of wetland. If the cost of restoration by a banker is lower than that by a developer, an offset scheme can achieve the efficient preservation of biodiversity.

As far as we know, however, there are few studies that examine offset schemes experimentally. Therefore, it has not yet been verified whether offset schemes – such as wetland-mitigation banking schemes in the United States – work as well as intended. Theoretically, when the area for preservation is determined appropriately, offset schemes can achieve the preservation of biodiversity efficiently and, accordingly, maximize social welfare. In reality, however, the introduced scheme may not work properly as expected due to the trading mechanism, characteristics of conservation targets, and the behavioral change of economic actors involved in trading. Here, imitating the real decision-making process and trading methods of economic actors, we evaluated the current offset scheme.

In the laboratory experiment, we assumed the existence of two regions. The environmental traders could perform credit transactions across regions, while the developers and bankers could only perform transactions within a region. Then, we examined whether the traders could improve the efficient creation of credits in both regions.

According to the results, the traders could not make a sufficient number of transactions to achieve efficiency in the first few periods. However, as they went through periods, their behavior became rational because of the learning-by-doing effect; therefore, the efficiency of credit creation was improved. The estimation result also revealed that environmental traders took into consideration the difference in the environmental values of both regions when they made transactions.

In the United States, which has a long history of biodiversity offsets, two types of banking systems have been introduced: one is mitigation banking, aimed at the offset of impacts on wetlands; the other is conservation banking, targeting the impacts on endangered species and their habitats. Similarly, Australia has constructed and managed two banking systems: one is bush broker, focusing on natural vegetation; the other is bio-banking, dealing with endangered species and ecosystems. From these experiences, we found the following two factors critical for the success of the biodiversity-offset banking system: long-term legal protection of the bank site not to be converted for development; and construction of the system providing management costs in the long term by means of establishing funds. Meanwhile, measures such as designation of small lands owned by individual holders as bank sites, and the creation of biodiversity utilizing the rights such that the other could use agriculturally abandoned lands would have many implications in Japan (Ota *et al.* 2011).

Ecosystem restoration and resource management

In addition to ecosystem conservation, its restoration was also analyzed. Here, defining return as a cost–benefit ratio and risk as an over-budget possibility, our project portfolio analysis demonstrated that international public agencies or environmental bodies could make better decisions on investment allocation to ecosystem restoration by applying project-portfolio viewpoints. The traditional approach for investment allocation regarding ecosystems has certainly emphasized ecological importance, but it also demonstrated that investment judgment from socio-economic aspects is superior to others in terms not only of socio-economic but also of ecological cost effectiveness. Briefly, considering socio-economic benefits and applying project portfolios are crucial for investment judgment on ecosystem restoration.

Focusing on fish species, we also performed economic analysis for resource management. Comparison and identification of the price volatilities among various fish species consumed in Japan was conducted to provide a better understanding of the price risks in the Japanese fish market. We find that the fish species whose market price had a relatively higher variance compared to other fish species were all protected under the Total Allowable Catch (TAC). This may be telling us that the current risky situations of Japanese fisheries are reflected in some of the Japanese retail fish markets. The price relationships among the retail fish markets of the ten highest-population cities in Japan are also analyzed and we configured how the Japanese regional fish markets can be delineated. Our results indicated that some fish markets like *aji* and *iwashi* are affected by regional factors, so that considering the characteristics of differences among different regional markets is important for conducting effective policy to protect the biodiversity of these resources in Japan.

Towards biodiversity conservation

Conventionally, policy measures for nature protection have focused on regulations and tax measures such as the designation of protected areas, regulation of development on lands and buildings, regulation of the harvesting and trading of rare species, and tax exemption for land owners within protected areas. However, regulation-oriented approaches have faced the difficulties of effective and efficient prevention of ecosystems and habitats degradation, resulting, in reality, in the destruction and separation of natural habitats. With this background, economic instruments such as the economic valuation of biodiversity and ecosystem services, biodiversity offset, PES, and eco-labeling for agricultural products are currently being introduced mainly in Western countries.

This book introduced how to perform an economic valuation of biodiversity and ecosystem services and to realize conservation which may provide more benefits to society than destruction by means of economic measures. Further studies on implementation of such measures, including the offset market, are expected.

References

Ota, T., Ito, H., Hayashi, K. and Malhotra, K., 2011, "Beikoku to goshu no seibutsu tayousei ohusetto bankingu sisutemu no hikaku (Comparison of biodiversity offset banking system between USA and Australia)," in S. Managi (ed.) and IGES, *Seibutsu tayousei no keizaigaku* (Economics of biodiversity), Kyoto, Japan: Shouwado.

Index